# THE FLIP

Epiphanies *of* Mind
*and the* Future
*of* Knowledge

*JEFFREY J. KRIPAL*

BELLEVUE LITERARY PRESS
*New York*

First published in the United States in 2019 by Bellevue Literary Press, New York

For information, contact:
Bellevue Literary Press
90 Broad Street
Suite 2100
New York, NY 10004
www.blpress.org

Library of Congress Cataloging-in-Publication Data
Names: Kripal, Jeffrey J. (Jeffrey John).
Title: The flip : epiphanies of mind and the future of knowledge / Jeffrey J. Kripal.
Description: First edition. | New York : Bellevue Literary Press, 2019. |
Includes bibliographical references.
Identifiers: LCCN 2018021661 (print) | LCCN 2018038244 (ebook) |
ISBN 9781942658535 (ebook) | ISBN 9781942658528 (trade paperback)
Subjects: LCSH: Religion and science. | Spirituality.
Classification: LCC BL240.3 (ebook) | LCC BL240.3 .K74 2018 (print) |
DDC 128/.2--dc23
LC record available at https://lccn.loc.gov/2018021661

Bellevue Literary Press would like to thank all its generous donors—individuals
and foundations—for their support.

Book design and composition by Mulberry Tree Press, Inc.

Bellevue Literary Press is committed to ecological stewardship in our book production
practices, working to reduce our impact on the natural environment.

∞ This book is printed on acid-free paper.

Manufactured in the United States of America.
First Edition

5 7 9 8 6 4

paperback ISBN: 978-1-942658-52-8
ebook ISBN: 978-1-942658-53-5

*An era can be considered over
when its basic illusions have been exhausted.*

—ARTHUR MILLER

# Contents

# THE FLIP

# THE HUMAN COSMOS

*Hydrogen is a light odorless gas that,*
*given enough time, turns into people.*

—ANONYMOUS

This little book is about many things. It is a report on the state of knowledge about the nature of mind and its relationship to matter, including the matter of the brain. It is an ironic, affectionate observation about how much visionary literature the scientific and medical professions have helped produce over the last century (think of the literature on the philosophical implications of quantum mechanics, on the near-death experience, on savant phenomena, and on psychedelic molecules). It is also a designed polemic and public complaint about the dangerous disregard for the humanities in contemporary culture, academic and otherwise. In the end, though, it is mostly an inappropriately hopeful, if not wildly optimistic, essay about a tipping point, about the future—be it near or far—of a new worldview, *a new real* that is presently forming around the epiphany of mind as an irreducible dimension or substrate of the natural world, indeed of the entire cosmos, before and beyond any present scientific, ethnic, political, or religious story that one happens to find one-self (caught) in at the moment.

And I do mean *epiphany* of mind. Among other rhetorical strategies, I mean to highlight and comment on a few examples of a large, scattered, but remarkably consistent set of stories about extreme life-changing experiences that intellectuals, scientists, and medical professionals have been reporting for centuries now but have written about with increasing visibility and effect only over the last few decades.

As these stories dramatically demonstrate, a radically new real can appear with the simplest of "flips," or reversals of perspective, roughly, from "the outside" of things to "the inside" of things, from "the object" to "the subject." And this can occur *without surrendering an iota of our remarkable scientific and medical knowledge about the material world and the human body.* The general materialistic framework of the sciences at the moment is not wrong. It is simply half-right. We know that mind is mattered. What these stories suggest is that matter is also minded, that this mindedness is fundamental to the cosmos, not some tangential, accidental, or recent emergent property of matter. Some of the stories go even further and suggest that matter may, in fact, be an expression of some kind of cosmic Mind, which expresses itself as the material world through the abstract structures of mathematics and physics.

What these stories also suggest is that abstract third-person knowledge or philosophical modeling of such mind is seldom, if ever, sufficient. At least in our present moment, it usually takes a deeply personal and direct encounter with this minded cosmos to convince an intellectual or scientist. That moment of realization beyond all linear thought, beyond all language, beyond all belief, is what I call "the flip." It is a *very* big deal. Such a flip is often sudden, unbidden, or traumatically catalyzed. It is also beautifully, elegantly simple.

The relative brevity of this book is meant to signal this simple beauty. I fully understand that most of my readers will never have experienced such a flip. The book, then, works primarily on the level of the intellectual imagination. It does not rely on or require such a direct experience. The pages attempt to flip the reader via story, philosophical argument, and simple human trust (in the otherwise-unbelievable stories that other human beings tell us here), as the pages tease out what exactly this flip is, what it might imply about us and our world, why it is so convincing to those who have known it (and so unconvincing to those who have not), how it reintroduces real meaning back into the real world, and, finally, what its moral, political, and spiritual implications might yet turn out to be. The book attempts all of this through five chapters that can be read in a single day. I do not want to exhaust you with words. I want to flip you.

Each chapter is an *essai*, as in the French—that is, a "test," an "attempt," or an "experiment" rather than a proclamation of certainties or a statement of settled convictions. I possess none of the latter. These are public thought experiments, professional complaints, moral struggles, friendly jabs, a thinking-through.

I begin by employing a set of common extraordinary experiences to call for a new recalibration of the humanities and the sciences toward some future form of knowledge.[1] That new coordination, I suggest, will come as contemporary neuroscience continues to fail, spectacularly, to explain consciousness through any materialistic model or causal mechanism and a new philosophy of mind begins to appear that understands consciousness (which is not to say ego, personality, or social self) as prior and primary and so irreducible to brain function or any other material mechanism.

With this irreducibility of mind will come the new ascent of the humanities, which, after all, have always been about engaging and interpreting both the most banal and the most fantastic ways that consciousness is reflected and refracted through the cultural codes of human civilization—that is, through history, social practice, language, art, religion, literature, institution, law, thought, and, I dare add, science.

I do not just focus on extraordinary experience, though. I focus on the extraordinary experiences of scientists, medical professionals, engineers, computer scientists, and highly trained intellectuals, including some Nobel laureates, each of whom flipped his or her previous materialistic outlook after experiencing some overwhelming event that revealed the fundamental, irreducible nature of mind as such. With such a professional focus, I want to shake the reader from the easy notion that such completely inexplicable events happen only to the naïve or to those who do not know their science. This assumption needs to be called out for what it is: utter and complete nonsense. The hyperreality and burning implications of these events for those to whom they happened have absolutely *nothing* to do with a rejection of science. Quite the opposite: They often result in new scientific ideas and even new technologies. Whatever they are (or are not), such flips appear to be scripted as goads and inspirations, not as blocks or trips. They appear to be pointing us to the new real and to the future of knowledge.

I do not leave it there. I attempt to hone the conversation further by seeking to interpret these epiphanies of mind through the apparent relationship of mind and matter implied or revealed in these types of experiences. Here we move from *what* the flip looks like in the lives of professional intellectuals to *how* it might work, or, to put it more technically, we move into a discussion of the ontological conditions (that is, the nature of the real itself) that would render the flip both possible and plausible, and—and this is a big one—that makes mathematical discovery and scientific knowledge possible at all.

Here I make the jarring but perfectly accurate observation that mind or consciousness is the subject and locus of *all* scientific practice and knowledge; that science, at the end of the day, is a function of human subjectivity and consciousness and not, as is often assumed, a simple photographic record of the world of things and objects "out there." If, however, science is finally an expression of human subjectivity, and if the same human sciences have been uncannily successful in peering into some of the deepest secrets of the universe, then the human subject itself must bear some intimate relationship to that same universe. The astonishing successes of science and the unreasonable ability of abstract mathematics to model and mirror the furthest reaches and cosmic history of matter, I suggest, are the best evidence for our own secret nature. *Human science works because human nature is cosmic.*

Part of how the flip works involves the dramatic and powerful ways that the event communicates meaning to the individual, often through baroque or fantastic imagery (think of the wild imagery of a near-death journey to "another world" or of a life-changing psychedelic "trip"). Conventionally, these images and narratives have been interpreted in entirely subjectivist or hallucinatory ways as fundamentally meaningless—that is, as possessing no real connection to the real world. That is a serious mistake, I will explain, and one that is easily avoidable once we distinguish between conventional and symbolic forms of communication and representation.

Finally, I explore some of the moral, social, and political implications of the flip. Not the *what* or the *how* of the flip now, but the *where to* and the *what for*. The single big idea here is that once one makes the flip and begins to understand that consciousness is fundamental, is a primitive of the physics and mathematics of the universe, it becomes more than apparent that every local religious ego or political identity, every local story, is historically relative, built on and constructed out of this deeper-minded matter or conscious cosmos. One can still affirm and nurture all of those local relative identities after the flip as intimate expressions of consciousness (and so one can also continue to act from within a particular story and its script, if one so chooses), but one will no longer make the dangerous mistake of privileging one's own inherited story and script over every other. One will recognize that there are many stories, many ways of enacting a form of reality, and that each of these do different things well (and other things poorly).

It really matters, then, which story one lives in (depending on what one wants to do well), but no story, however "sacred" or "scientific," can or ever will be absolute and speak for all of human experience and human potential, much less all of earthly or cosmic life. This is not a curse. This is a promise, a gift, and a preservation. As in biological evolution, so, too, in human culture and consciousness: Pluralism and diversity are precious goods that enable life to survive, flourish, and experiment, like an artist at work.

The flip, in short, relativizes and affirms each and every culture, community, and religion, even as it cosmicizes and—I dare say—spiritualizes our shared humanity. The flip results in a new *cosmic comparative* perspective that reorients us within an immeasurably larger vision of who we are as a species of the cosmos and what we might yet become. The future of knowledge, it turns out, is also the future of us.

• • •

I recognize, of course, how far we are away from any and all of this. I am not naïve. I have suffered serious harassment and backlash over the years for my writings on this cosmic humanism both from politically motivated religious censors (for my insistence on the sexual—read material—dimensions of many forms of ecstatic religious experience) and, to a much lesser extent, from secular critics (for my insistence on the paranormal or mind-over-matter dimensions of American culture and history). I know perfectly well that the form of mind I inhabit, the seemingly paradoxical "third way" of the flip—at once deeply critical of and deeply sympathetic to all local religious expressions—is very difficult, impossible really, for both the religious fundamentalist and the ideological materialist to understand and accept (and, as I will show, these two mind-sets have much more in common than either wants to admit).

I nevertheless think that this third way "beyond belief" and "beyond reason" is far preferable to religious belief or pure mechanistic rationalism, since it opens up new horizons of inquiry and thought and does not prematurely shut down our quest for meaning, which is exactly what belief and hyperrationalism do in different ways.[2] Put a bit differently, I am convinced that this third way represents our best way forward into the future, into new ways of knowing and new conceptions of the human that we have only imagined at this point in genres like comparative mystical literature and science fiction. I do not think these future forms of knowledge will be "religious" in the traditional sense, any more than I think they will be "scientific" in the classical materialist sense. They will be both, and neither. They will be something else, and something way, way more.

The flip, then, gives us a new way of revisioning and renewing the humanities in deep conversation with the sciences. It enables us to tell a richer and, frankly, more accurate history of science and medicine (since such a history has been informed and inspired by anomalous or "flipped" states of mind from the beginning). The flip points to new ontologies and epistemologies on the horizon of thought. And it suggests a new philosophical, really cosmic, foundation for a future ethics and politics. Obviously, this is not a minor project. Nor is it a humble or cautious one. This is a manifesto. Short. Irreverent. Punchy. Blunt.

And why not? Do we really have that much time for self-censoring politeness and endless qualifications, which are too often also obfuscations? I don't think so. *The Flip* is an intervention into our present fraught political moment—fraught because we appear to have lost any sense of the cosmic human and have shrunk ourselves down to this or that minuscule religious, nationalist, secular, ethnic, or genetic ego. We are shrinking into oblivion. We have it all exactly upside down. We have forgotten, or not yet realized, our own secret giant grandeur. And so we suffer.

May you not suffer like this any longer. May the present "you" not survive this little book. May you be flipped in dramatic or quiet ways.

# 1

# VISIONS OF THE IMPOSSIBLE

*But, as always, the key to making sense of our lives lies in those details that seem most nonsensical. The small strangenesses surrounding us are our best possible clues to reality.*

—PETER KINGSLEY, *Reality*

Two impossible true tales.

Scene 1. Twain's Mental Telegraphy. Dressed in his famous white "dontcaredam suit," Mark Twain was famous for mocking every orthodoxy and convention, including, it turns out, the conventions of space and time. As he related the events in his diaries, Twain and his brother Henry were working on the riverboat *Pennsylvania* in June 1858. While they were lying in port in St. Louis, the writer had a most remarkable dream:

> In the morning, when I awoke I had been dreaming, and the dream was so vivid, so like reality, that it deceived me, and I thought it *was* real. In the dream I had seen Henry a corpse. He lay in a metallic burial case. He was dressed in a suit of

> my clothing, and on his breast lay a great bou-
> quet of flowers, mainly white roses, with a red
> rose in the centre.

Twain awoke, got dressed, and prepared to go view the casket. He was walking to the house where he thought the casket lay before he realized "that there was nothing real about this—it was only a dream."

Alas, it was not. A few weeks later, Henry was badly burned in a boiler explosion and then accidentally killed when some young doctors gave him a huge overdose of opium for the pain. Normally, the dead were buried in a simple pine coffin, but some women had raised sixty dollars to put Henry in a special metal one. Twain explained what happened next:

> When I came back and entered the dead-room Henry lay in that open case, and he was dressed in a suit of my clothing. He had borrowed it without my knowl-edge during our last sojourn in St. Louis; and I recog-nized instantly that my dream of several weeks before was here exactly reproduced, so far as these details went—and I think I missed one detail; but that one was immediately supplied, for just then an elderly lady entered the place with a large bouquet consisting mainly of white roses, and in the centre of it was a red rose, and she laid it on his breast.[1]

Now who of us would not be permanently marked, at once inspired and haunted, by such a series of events? Who of us, if this were *our* dream and *our* brother, could honestly dismiss it all as a series of coincidences? Twain certainly could not. He was obsessed with such moments in his life, of which there were all too many. In 1878, he described some of them in an essay and even theorized how they work. But he could not bring himself to publish it, as he feared "the public would treat the thing as a joke whereas I was in earnest." Finally, Twain gave in, allowed his name to be attached to his own experiences and ideas, and published this material in *Harper's* magazine in two separate installments: "Mental Telegraphy: A Manuscript with a History" (1891) and "Mental Telegraphy Again" (1895).[2]

Mental telegraphy. The metaphor points to the cutting-edge technology of the day. It also points to Twain's conviction that such precognitive dreams and instant communications were connected to the acts of reading and writing. Indeed, Twain suspected that whatever processes this mental telegraphy named had some profound relationship to the deeper sources of his own literary success. And he meant this quite seriously. The "manuscript with a history" of the first essay title refers to a detailed plotline for a story about some Nevada silver mines that came blazing into his mind one day, as if out of nowhere, as if from someone else. When a letter from a friend three thousand miles away arrived in the mail a few days later, he knew exactly what was in it before he opened the envelope: the plot of the silver mine story that he had received in a flash of creativity and inspiration a few days earlier.

Scene 2. The Wife Who Knew. Then there is the American forensic pathologist Dr. Janis Amatuzio. Her book *Beyond Knowing* is filled with extraordinary stories of impossible things that routinely happen around death. Here is one such true tale.

This one began one night when Amatuzio encountered a very troubled hospital chaplain in the course of her work. He asked to go back to her office, where he then asked her if she knew how they had found the body of a young man recently killed in a car accident. She replied that her records showed that the Coon

Rapids Police Department had recovered the body in a frozen creek bed at 4:45 A.M.

"No," the man replied, "Do you know how they *really* found him?" The chaplain then explained how he had spoken to the dead man's wife, who related how she had had a vivid dream that night of her husband standing next to her bed, apologizing and explaining that he had been in a car accident, and that his car was in a ditch, where it could not be seen from the road. She awoke immediately, at 4:20 A.M., and called the police to tell them that her husband had been in a car accident not far from their home, and that his car was in a ravine that could not be seen from the road. They recovered the body twenty-five minutes later.[3]

### From the Preternatural to the Paranormal

Impossible, right?

We have no idea what to do with such poignant, powerful stories. So we disempower them with words like *anecdote* and *coincidence*. Or perhaps we could study their textual histories and show that they were not really this clean or simple. That would be a relief. Like the heads of Hercules' Lernaean Hydra, however, with every story we so decapitate, three more, or three thousand more, would appear. We are, in fact, swimming in a sea of such stories at this very moment, if only we could recognize our situation and its strange signs.

We cannot recognize our situation because we have shamed every category and every word that might help us. Consider that most difficult word *paranormal*. Most journalists, scientists, and even intellectual historians (who should know better) demonstrate little more than total ignorance of the word's philosophically nuanced origins and its actual place in the history of science.

It appears that the word was coined by the French researcher

Joseph Maxwell in 1903. Maxwell was no naïf. He was a prose-cuting attorney who became the president of the court of appeals of Bordeaux. He was also a medical doctor with an advanced degree and dissertation on amnesia and disorders of con-sciousness in epileptics. In his book *Les Phénomènes psychiques: recherches, observations, méthodes,* he used the term *paranormal* to describe mind-over-matter phenomena that, already then, were very well documented, if poorly understood. Think telekinetic, poltergeist, or materialization phenomena. The *paranormal*—literally "to the side of" or "beyond" (*para-*) the normal—was almost certainly a French gloss on an earlier English word, this one coined by a Cambridge-trained classicist and education reformer, Frederic Myers: the *supernormal.*

It is crucial to understand that both the English and French adjectives did not imply or require anything supernatural or miraculous—that is, "from God" or outside the natural world (although they did not exclude such possibilities, either). Rather, both words were coined to describe our own almost total igno-rance of all of those fantastic phenomena that are a part of our human nature and the natural world but that we cannot yet model or explain within any adequate scientific framework.[4]

These new words, then, were not naïve expressions of credulity. Nor were they a mark of some willful ignorance of science. They were coined and used by some of the most educated minds of Europe to explore anomalous phenomena that appeared to signal some richer reality than the present science could explain but that, it was hoped, some future science would. Both words were humble and honest placeholders, markers of a deep intellectual humility and a radical empiricism that refused to look away from things it could not explain and saw anomalies not as idiocies but as mean-ingful signs pointing toward some future form of knowledge, some *new real*. So, too, here. I do not want to overemphasize or fetishize such phenomena. I simply want to call out those who want to claim that they do not happen. They do. I also want to suggest that such strange signs could guide us on our way, if only we would listen and look and not turn away.

Really, all that is required is a single space. The supernormal and the paranormal did not mean the supernatural, which functioned from the thirteenth century on as a clear marker for an act or event "from God"—that is, as a miracle that issued from outside the natural order. The supernormal and the paranormal meant, rather, to refer to all of those strange signs and anomalies of human experience that hint at a fundamental reality that is, yes, well "above" or "beyond" (*super*) our present scientific *or* religious models but that nevertheless remain very much a part of a single nature, now conceived in much more expansive and fantastic terms. Such new words signaled the super natural.[5]

There were, of course, numerous earlier categories that attempted the same basic move. One was the Latin-based adjective *preternatural*, which was used in natural philosophy to refer to marvels and anomalies that seemed inexplicable but that the authors were reluctant to trace back to the work of God and so could not be called "supernatural." In effect, the preternatural sat between the natural and the supernatural as a third mediating category.

The skeptical reader at this point will no doubt dismiss all of these categories as unfortunate results of "superstitious" thinking—religious nonsense that real intellectuals always debunk and that we have, thankfully, moved beyond.

Good luck with that. The real history of science and religion works strongly against such easy assumptions, which are frankly ignorant. There are endless examples. Take the sixteenth-century Latin expression "natural divination" (*divinatio naturalis*), as it has been traced by the historian of science Andreas Sommer. The phrase meant more or less what the early twentieth-century coiners of the paranormal meant: entirely natural, if rarely expressed, capacities of the human organism, here linked to what today we would call precognition but which earlier cultures knew as "divination" and "prophecy."

Obviously, this was another unfortunate superstition by a poorly educated fool. Except that the expression was positively used by none other than Francis Bacon (1561–1626), the English aristocrat widely considered to be the father of modern empiricism, inductive reason, and the scientific method. Unlike his modern disciples, Bacon did not make the mistakes of denying reality to phenomena that he could not explain or, more subtly, restricting all human knowing to ordinary states of cognition, rationalism, and sensory input.

Indeed, he wrote of not one, but two forms of natural divination: One he called "primitive," which was restricted to the human person, and one he understood as a result of "influxion"—that is, inspiration from disembodied spirits. He linked the former to altered states of concentration, particularly dreams, religious ecstasy, and being "near death," in ways that uncannily look forward to the modern near-death literature, in which such "near-death experiences," as we call them now, commonly catalyze exactly these same human potentials. According to him, primitive natural divination occurs when "the mind, when it is withdrawn and collected into itself, and not diffused into the organs of the body, hath some extent and latitude of prenotion, which therefore appeareth most in sleep, in extasies, and near death, and more rarely in waking apprehensions."[6]

The standard historical narrative of some kind of inevitable "scientific progress" in which scientific knowledge marches on and eliminates one such notion (or prenotion) after another until we inevitably get to our present-day materialist orthodoxy is simply not true. It is pure ideology. Convictions and phenomena like Bacon's natural divination were never eliminated. They were simply ignored. Moreover, Bacon was hardly alone in his occult convictions among the giants of modern science. Galileo read horoscopes for his patrons, his daughters, and himself. Newton wrote more about occult and alchemical matters than about mechanics and mathematics.[7] Albert Einstein wrote an appreciative preface for a book on telepathy (Upton Sinclair's 1930 *Mental Radio*). And the same patterns continue right down to today among elite scientists and medical professionals.

### Paranormal Criticism

Contrary to what is assumed, such anomalous events do not automatically lead to belief, much less to organized religion. Sometimes, in fact, they lead *out of* belief and *out of* religion but also *into* something other and more. The concept of the paranormal, then, is inherently "bimodal."[8] It is a perfect expression of what I have called the third way of the flip, at once radically critical and deeply sympathetic. As such, it points back to the experiential origins of much religious belief and ideation (the soul, immortality, mind-over-matter or "magical" powers, and so on), and yet it also suggests deeper, natural or super natural processes behind or within all of these allegedly "supernatural" events. As such, the paranormal is at once disenchanting and enchanting, both deeply suspicious of all supernatural explanations but also open to new super natural ones.

I once gave a lecture to a group of Orthodox rabbis and Jewish academics on some carefully documented cases of dream precognition (think time-stamped e-mail trails) by a contemporary

Jewish woman with whom I have worked closely.[9] One of the respondents to my paper articulated a vague challenge that amounted to the claim that the paranormal is "against religion." If I might intuit and translate his concern, I think what he meant is that the paranormal phenomena I was describing did not depend on any particular religious tradition or revelation; that what I was really describing were universal human potentials that are independent from religious tradition as such.

I did not disagree with this concerned colleague, although I tried to articulate how the same phenomena lie at the origins of religious revelation, and that they need not be "antireligious," either. Sometimes it is disenchanting (as it was for him). Sometimes it is enchanting (as it was for the woman I was lecturing about). It all depends on how it is interpreted and picked up by an individual or community. Still, I think he was onto something, and I was grateful for both the observation and the challenge.

I think the most basic and uncontroversial thing we can say about paranormal experiences is that they appear to lie below or behind basic religious ideas and practices, such as the separable soul, immortality, reincarnation, transcendence, and divination, to name a few. We do not need to accept the ontological reality or empirical objective truth of any of these ideas to see that (a) these types of sincere and honest experiences are as evident, really *more* evident, in the present as/than they are in the irrecoverable past; and (b) they naturally produce in their subjects a deep and abiding conviction in the truth of specific religious ideas. We do not need to believe the beliefs to understand and appreciate their basic rationality or reasonableness. Most of us, after all, would, in fact, at least entertain the idea of a separable immortal soul if we experienced ourselves separating from our bodies and floating above a car crash or surgery room. I certainly would.

This does *not* mean the beliefs are true, only that they are perfectly reasonable and understandable. Put in the terms of the contemporary study of religion, we might say that such beliefs are not simply constructions, representations, or products of the material transmission of texts and doctrines. Rather, such beliefs appear to arise from universal human capacities or uncanny potentials that are then, of course, shaped, constructed, textualized, and transmitted by all of the processes that scholars of religion have come to know so well.

The scholar of religion Ann Taves has referred to extraordinary religious experiences, what she calls "revelatory events," as the "building blocks" of religion. They are not yet "religion," nor need they become "religious." They must be construed or interpreted religiously to become so. If these same religiously construed extraordinary events are then taken up and transformed by the various social, political, cognitive, narrative, and textual processes that are the present concern of the study of religion, they might well be "built up" into religious forms and institutions.[10] We might thus start with a series of postmortem physical or photic apparitions (the Resurrection appearances remembered in the Gospels and Paul's famous speaking light on the road to Damascus) and end up with something like Christianity. What the historian of religions literate in the psychical research tradition knows that the believer or even scholarly colleague does not generally know is that such postmortem apparitions, even the apparently physical ones, are surprisingly common in human history, but that only a tiny few get picked up and are used as building blocks for new "religions."

There is a hidden gift, and a hidden hit, in this building-block approach. After all, if extraordinary experiences or paranormal events constitute the building blocks of religion, if they are *proto*religion, then any religious tradition can be taken apart and reduced to these protoforms or building blocks. The metaphor of the building block implies a reduction to a more basic unit.

I think this is correct, and I think it is why paranormal experiences and experiencers are often imbued with robust forms of criticism vis-à-vis "religion," but of a very particular double kind. Such experiencers often (not always) recognize *both* the truth of the experiences themselves *and* the arbitrariness of the way these experiences are built up into public religious forms. This, in turn, I suspect, is what makes paranormal experiencers and experiences so culturally creative. It is also what makes them so often resisted or feared. Here is the perfectly accurate (but incomplete) intuition that my Jewish colleague was trying to articulate after my talk to the Orthodox rabbis when he told me that the paranormal is "antireligious."

The truth is that, far from always being the wide-eyed, gullible fools that the media likes to portray, these experiencers often operate with extremely sophisticated forms of self-criticism and professional expertise. Perhaps not accidentally, these experiencers are often professional intellectuals, gifted writers, and world-class scientists.

In the fall of 2016, I spent three days discussing these matters with a number of accomplished scientists from a variety of fields, from rocket science to information science. All were convinced that the phenomena are real and deserve major attention, but none construed them in traditional or religious ways. One of these individuals, Kary Mullis, won the Nobel Prize for chemistry in 1993 for his invention of the polymerase chain reaction. Kary spoke forcefully about a series of alienlike abduction events on his northern California cabin property that involved himself, his daughter, and another scientific colleague. All three individuals independently experienced an eerily similar presence on the same cabin property over a number of years. Kary was "gone" for six hours. His daughter was "gone" for three hours, as her fiancé frantically searched and screamed for her. Both father and daughter "woke up" walking down the same road toward the cabin, completely unaware of the previous lost hours or what had happened to them.

Mullis resists the sci-fi interpretation that this abducting presence was some kind of "alien." That is an overinterpretation for him. But he absolutely refuses to deny what happened to him, indeed what happened to all three of them, independently and on the same property.[11]

*That* is the double move I want to underline here as "paranormal criticism." It is time to affirm the historical reality of these events without signing our names to any particular mythological or religious framing of them. The results, I hasten to add, are not uncritical with respect to religious belief. Quite the contrary, they can be devastatingly critical, far more critical than any purely secular approach could dream of being.

Consider two of the most secular public cultures on the planet: those of Sweden and Denmark. Consider also the secular theorist Phil Zuckerman, who published a book on these two cultures a few years ago entitled *Society Without God*. Significantly, Zuckerman's secular subjects consistently report dramatic paranormal experiences.[12] Unlike religious people, however, the Scandinavians do not generally use them as means to build up or support a religious worldview—that is, they do not construct public belief systems out of them. They simply let them stand. But they do not deny them, and they fully recognize how incredibly strange they are. To Zuckerman's great credit, he does not deny them, either, and this despite the fact that he is one of our most accomplished secular theorists in the study of religion. He discusses them openly as common human experiences that are obviously sincere and important. His intellectual generosity and basic human fairness is simply remarkable, and, unfortunately, all too rare in these waters.

What is so significant for our present purposes, though, is how these Nordic paranormal events function as robust forms of cultural and religious criticism when they do function in a religious context or interpretive framework. Experiencers here do not generally construct or "believe." They deconstruct and "take apart." Consider one of Zuckerman's star subjects: Johanne. Johanne is a Danish novelist, painter, and religious studies Ph.D. student who is married to a Lutheran minister. Unlike most Danish citizens, Johanne believes in God, although her belief is hardly a traditional one. She told Zuckerman two separate paranormal stories, I assume, to help explain her nuanced relationship to what she means by "God" and "religion."

One involved a detailed "feeling" or profound fear concerning her two-year-old daughter, Eva. Here is a classic case of "telepathic" communication between two entangled loved ones, as we saw earlier with the dead husband appearing to the living wife. Here, though, the communication does not just violate what we think of as space. It violates what we think of as time. Eva was scheduled to go on a field trip to Aarhus to see Santa. She was to travel on a train the next day. Johanne fought her fear all the previous day. While giving Eva a bath, she thought she saw her lying on some train tracks in front of a train—obviously, a very upsetting image for a parent. She did not want to let her daughter go the next day, but her husband, Mikkel, won the argument, and she took Eva anyway, almost turning back home twice on the way to the school. Johanne waited all day for the phone to ring. It did not ring, but when Mikkel returned from the school with Eva, he had something very upsetting to tell her. Eva had gone missing and was found lying on the train tracks in front of the train, exactly as she had foreseen. Luckily, the train was not moving and no harm came to the child.[13] Now here is the second story, with a punch. Johanne and Mikkel were driving in a car and discussing a television program that Mikkel was to appear on to give a brief sermon. During the conversation, Johanne grew very disillusioned: "I don't want to discuss this old stupid stuff written 2000 years ago," she finally said in exasperation. And then it just happened, in the car, right then and there. Johanne breaks the frame of the story and apologizes to Phil Zuckerman: "I'm sorry, I know this must sound absolutely idiotic to you, but . . ." And then she explains what happened next:

> Then suddenly I just started to speak without intending to. And then I got like . . . I think I called it to myself like a revelation, . . . I just started to speak and I was seeing stuff. I saw like a corpse passing by and I saw the faculty of theology burning and disappearing into the sky, . . . I kept talking in a very fluent way, very . . . like I was not talking myself. And then I saw Martin Luther and his whole—what you call it, the Reformation and—and then I started talking about this Reformation, that this was not enough. We had to get back to something even more fundamental than just . . . yeah, just making a revolution against the Catholic Church and we had to get back into even more fundamental—and then—it's all that we know in the Bible, everything is just bullshit, everything is bullshit. I didn't say bullshit, but I said it in a very poetic language, and then suddenly I just saw

all these corpses in the sky and—in a very split second, it was like everything stopped and I just saw just a glimpse of something. And then I said to Mikkel, and all this is just nothing—the real stuff is something completely different.[14]

Much could be said at this point about the context and content of Johanne's channeling and vision, including the obvious influence of her artistic and intellectual training in the study of religion (the corpses of "the faculty of theology burning and disappearing into the sky"). Suffice it to say that the total experience was a profoundly deconstructive one, but please note that it was a deconstruction toward a positive "something" so profound that she is willing to call it a "revelation." We are never told what this "real stuff" or "something" is (apparently, she honestly does not know), only that Johanne was given a glimpse of it, and that it is "totally different" from what the Bible says. That's all old stupid stuff. That's all bullshit.

### The Rules of the Academic Table

We do not know how many such stories there might be or, much less, what they might mean. We do not know because we have never really tried to find out. Why, after all, would we invest in a study of something that does not exist?

"Water?" the fish asks. "What's water?"

It is worse than that, though. It is not just that we are told that such things that happen all the time cannot happen at all. It is that there are subtle, and not so subtle, punishments in place for those who take such events seriously—that is, for those who let the Hydra stand. Eyes roll. Cruel things are said. People retreat into safe silence.

Note that the first two of our opening stories feature a kind of professional fear. Twain struggled for years with whether to own his experiences in print. Even the hospital chaplain, who, one would assume, was used to death and various languages of the soul, was shaken to the core by what he had encountered. He, too, was not ready, was not prepared *for this*. Clearly, these events violate something very basic about our worldview and our established ways of knowing. That is no doubt why Dr. Amatuzio entitled her book *Beyond Knowing*.

It is not just our fault, though. There are also fundamental ambiguities inherent in the experiences themselves, ambiguities that make it very difficult to put and keep them on our academic tables. These things are not things, for one. Nor are they replicable or measurable, not, at least, in their most robust forms.

There is also the key role that the human imagination plays in these visions. I have related a few more or less straightforward cases, but the records are filled with more difficult, and more symbolic or outright mythical, accounts, whose high strangeness would boggle even the most generous among us. And even the relatively simple and more empirical cases are often shifted in little ways (missing an important detail or supplying a nonexistent one), which suggest that the imagination is at once constructing and seeing these visions.

The early Victorian researchers called dreams like the two with which I began "veridical hallucinations." That seems about right. The expression was designed to signal the paradoxical truth that such cognitions were functions of the human imagination and so obviously dreamlike or hallucinatory, and yet they provided accurate empirical information about what had already happened in the physical world or, weirder still, what was about to happen.

We are not very good at such paradoxical ways of thinking

today. We tend to think of the imagined as the imaginary, as fancy and schlock, but clearly something else is shining through these extreme cases, something, well, *true*. Somehow, Twain's dreaming imagination knew that his brother would be dead in a few weeks—it even knew what kind of flower bouquet would sit on his brother's breathless chest. Similarly, the wife's dream-vision knew that her husband had just been killed and where his body lay. Words like *imagined* and *real*, *inside* and *outside*, *subject* and *object*, or *mental* and *material* cease to have much meaning in these moments. And yet these words name the most basic structures of our knowing.

Or not knowing.

Then there are the little problems of space and time. The first two stories are finally about a kind of traumatic transcendence—that is, a visionary warping of space and time effected by the gravity of intense human suffering. The mother's precognitive dream of her little daughter on the train tracks the next day can also easily be placed here. Even space and time, these most basic "categories of the understanding," as Immanuel Kant had it, surrender their reign before the needs of the human heart. Much as Kant argued, space and time appear to be *our* cognitive filters, not some perfect reflection of what is really "out there," or, I dare add, of what we are really capable of knowing.

There may have been more to Kant's fundamental insight than philosophical precision.[15] On August 10, 1763, the philosopher marveled (in a private letter) at the clairvoyant abilities of the Swedish scientist-seer Emanuel Swedenborg, who in the year 1756 related to some dinner guests at Gothenburg the precise details of a fire advancing in a southern suburb of Stockholm, fifty miles away. From 6:00 to 8:00 A.M., the seer reported on the fire's advance, until it was finally put out, just three doors from his own home. In the next few days, all of this was investigated and confirmed by the Stockholm Chamber of Commerce and the governor. But here is the catch: Kant may have clearly accepted in private the empirical accuracy of such extraordinary powers, but he mocked and made fun of Swedenborg in public print. There is that professional fear again.

It is the soon-to-be-dead brother, the deadly car accident, the child lying on the train tracks in the future, or the neighborhood fire piece—what we might call the "privileging of the extreme condition"—that the debunkers misunderstand when they ask, with a sneer, why all psychics do not get rich on the stock market, or why robust psychical phenomena cannot be made to appear in the controlled laboratory. Putting aside for the moment the fact that psychics sometimes *do* get rich and that humble statistical forms of psychical phenomena do, in fact, appear in laboratories all the time, the answer to why *robust* events like those of Twain, the widowed wife, the anxious mother, and the Swedish seer-scientist do not appear in the lab is simple enough: There is no trauma, love, or loss there. No one is in danger or dying. Your house is not on fire. Your little child is not sleeping on the train tracks.

The professional debunker's insistence, then, that the phenomena play by his rules and appear for all to see in a safe and sterile controlled laboratory is little more than a mark of his own serious ignorance of the nature of the phenomena in question. To play by these rules is like trying to study the stars at midday, and then claiming that they don't exist because they do not appear under those particular conditions. It is like going to the North Pole to study those legendary beasts called zebras.

Hence it is utterly unsurprising, indeed perfectly predictable, that the controlled laboratory evidence involves slight but statistically significant patterns rather than the dramatic, blow-your-mind, knock-your-socks-off moments, which is precisely what one commonly encounters in the real-world traumatic cases. I have called this basic pattern "the traumatic secret," but I am hardly alone here. The same fact has been noted by many others, including the prominent physicist Freeman Dyson.

Perhaps you will not, *cannot*, hear me because I am not a scientist, because I work in the humanities, which you assume are not "real" forms of knowledge. Okay. Listen to Dyson. His recounting of his own argument is important not only because he was for most of his career a celebrated scientist at an elite institution, the Institute for Advanced Studies in Princeton, but also because he uses the adjective *anecdotal* as it should be but seldom is used in these contexts: as a perfectly accurate descriptor of another word—*evidence*. Here is Dyson looking back on a book review he had written for *The New York Review of Books*:

> In my review I said that ESP only occurs, according to the anecdotal evidence, when a person is experiencing intense stress and strong emotions. Under the conditions of a controlled scientific experiment, intense stress and strong emotions are excluded; the person experiences intense boredom rather than excitement, and so the evidence for ESP disappears, . . . The experiment necessarily excludes the human emotions that make ESP possible.[16]

The generous wisdom of the physicist aside, it is precisely these latter traumatic and shocking cases that the professional debunkers whisk away with their magical wand of the "anecdotal" and so take off the table of serious discussion. This is nothing but a mark of ignorance, or of bad faith. I am not sure which.

This same privileging of the extreme, however, is employed in the sciences all the time. As Aldous Huxley pointed out long ago in his own defense of "mystical" experiences, we have no reason to think from our ordinary experience that water is composed of two gases fused together by invisible forces. We know this only by exposing water to extreme conditions, by "traumatizing" it, and then by detecting and measuring the gases with advanced technology that no ordinary person possesses or understands.[17] The situation is eerily analogous to impossible scenarios like those of Twain, the wife, and the mother. They are generally only available in extreme traumatic or dangerous situations.

Nothing in our everyday experience gives us any reason to suppose that matter is not material, that it is made up of bizarre forms of energy that violate, very much like spirit, all of our normal notions of space, time, and causality. Yet when we subject matter to exquisite technologies, like the Large Hadron Collider near Geneva, Switzerland, then we can see quite clearly that matter is not "material" at all. But—and this is the key—we can only get there through a great deal of physical violence, a violence so extreme and so precise that it cost billions of dollars, necessitated the participation of tens of thousands of professional physicists, mathematicians, and computer scientists, and required decades of preparation to inflict it and then analyze its results. Hence the recent discovery of the "God particle," or Higgs boson at CERN.

We invested our energies, time, and money there, and so we are finding out all sorts of astonishing things about the world in which we live and of which we are intimate expressions. But we will not invest them here, in the everyday astonishing experiences of human beings around the world, and so we continue to work with the most banal models of mind—materialist and mechanistic ones—that is, models that assume that "mind equals brain" and that the psyche works like, or *is*, a computer. What is going on here? Why are we so intent on ignoring precisely those bodies of evidence that suggest that, yes, of course, mind is correlated with brain, but it is not the same thing (and, oh, by the way, mind created the computer, not the other way around). Why are we so afraid of the likelihood that we are every bit as bizarre as the quantum world; that we possess fantastic capacities that we have so far only allowed ourselves to imagine in science fiction and fantasy literature?

Take my own discipline, the history of religions. It is filled with countless "impossible" things—from conscious light forms to materializing monsters to levitating saints—that make my opening stories look utterly ordinary and uneventful. And what do scholars of religion do with these astonishing (im)possibilities? Not much. We are told endlessly, and quite correctly, that religious experience of every sort is always "constructed" by local languages, ritual practices, and institutions. We thus insist on "contextualizing" every experience and event, which means locking them down tight to a particular physical point in space-time and so not allowing them to inform how we understand other obviously similar experiences and events at other points of space-time. Such "comparisons" are deeply suspect these days. They are called all sorts of bad names, partly, yes, because bad comparisons have historically been used to advance colonial projects (with so-called primitive religions "evolving" into exotic "Oriental" polytheisms and, eventually, into "advanced" civilization—that is, into European Christianity), but mostly, I think, because such comparative methods always end up suggesting something at work in history that is not strictly materialist or, frankly, historical. Like a man dreaming the future of his brother's funeral, or a departed soul-body showing up in the dream of a sleeping wife. Colonialism is bad, but *that*, apparently, is even worse.

In the same vein, we are also told (again, quite correctly) that religion is about power and politics, or economics, or patriarchy, or empire and colonial oppression, or psychological projection, or the denial of death, or—now the latest—cognitive templates, evolutionary adaptation, and computerlike synapses. And, ultimately, of course, what religion is really about is absolutely nothing, since we are nothing but meaningless, statistically organized matter bouncing around in empty dead space. In the rules of this classical materialist game, the scholar of religion can never take seriously what makes an experience or expression religious, since that would inevitably involve some truly fantastic vision of human nature and destiny, some transhuman or superhuman possibility, some mental telegraphy, dreamlike soul, clairvoyant seer, or cosmic consciousness. All of that is taken off the table, in principle, as inappropriate to the academic project. And then we are told that there is nothing "religious" about religion, which, of course, is perfectly true, since we have just taken all of the fantastic stuff off the table.

Put a bit differently, *our conclusions are really a function of our exclusions.* What we think the stuff on the table means is mostly a function of what we have taken off that table. In the language of the French philosopher Michel Foucault, our "order of knowledge" is highly "disciplined" and "policed." And the police are the classical materialists, be they scientists or humanists (for the humanities are nearly as policed by classical materialism as the sciences are). If, of course, we were to put the stuff in the materialist's wastebasket back on the table, what the table "means" would shift, and shift dramatically.

I am reminded of a classic sci-fi movie. We have become something like the protagonist Scott Carey at the very end of *The Incredible Shrinking Man* (1957). Even in the microscopic world, however,

Scott spiritually realizes that he still exists and means something. Not anymore. With every passing decade, human nature is getting tinier and tinier and less and less significant. In a few more years, we'll just blip out of existence and meaning altogether, reduced to nothing more than cognitive modules, replicating DNA, quantum-sensitive microtubules in the synapses of the brain, or whatever. Either that or these same methods will simply kill us off. Indeed, at this point, we in the humanities are constantly reminded of the "death of the subject" and told repeatedly that we are basically walking corpses with computers on top—in effect, technological zombies, moist robots, meat puppets. We are in the fantastically ridiculous situation that conscious intellectuals are telling us that consciousness does not really exist as such, that there is nothing to it except cognitive grids, software loops, and warm brain matter. If this were not so patently absurd, it would be very funny.

### Dead Minerals or Supernatureculture

I do not remember how old I was. Maybe eight? I was an avid reader of the *Encyclopedia Britannica*, which basically functioned as a kind of paper Internet in the early 1970s. One day, my elementary school class took a field trip to the local post office. The postmaster led us through the building, explaining this and that, until we eventually ended up in the basement, which happened to be one of the few nuclear shelters in this little midwestern town. One of those disturbing Cold War nuclear signs hung ominously near its door. We entered. The postmaster showed us metal casks of water and stacks of boxes of (no doubt stale) crackers that were stored there, presumably as drink and food for the town's inhabitants (that is, us) in case of a Soviet-U.S. nuclear exchange. Looking back, I am sure the stock was ridiculously inadequate, but that never quite entered our elementary heads.

Trying to explain the importance of the shelter, he asked, "What is a human life worth?" I immediately raised my hand. The poor postmaster called on me. "Four dollars and thirty-five cents," I confidently stated. (Or something like that.)

Though the postmaster was confused, everyone laughed. But I was being perfectly serious. And why not? I had just read in the *Encyclopedia Britannica* that if one gathered together all the minerals of the human body, this is what they would be worth. It never occurred to me that "worth" (being a function of a human economic system) was an entirely artificial construct. It also never occurred to me that I had just reduced the postmaster's "human being" to nothing but dead minerals. Or, for that matter, that the thought experiment I had enthusiastically repeated was an utterly gruesome one that implied a horror-movie scenario—harvesting minerals from a corpse.

The anthropologist Mayanthi L. Fernando recently wrote a most remarkable reflection entitled "Supernatureculture" for the online magazine of the Social Science Research Council—a revered and elite intellectual context, if ever there was one. The essay begins with her cat, Hoppy, knocking over a water glass of flowers near her bed. No big deal, except that Hoppy had been dead for four days. It was clearly Hoppy, though, engaging in her signature move in physical life.

Fernando spends the rest of the essay stretching every social scientific and physicalist category that we possess, including the quantum ones, to make sense of what was patently obvious to her: that an entirely invisible, immaterial nonhuman presence had intervened in her life, and quite playfully no less. The piece ends with two more little stories. Fernando describes telling about Hoppy in two different public lectures. In the first, the earlier presenter fainted at the podium. In the second, a glass of water spilled all over the table.

Here we encounter the recursive, reflexive, and openly mischievous nature of the paranormal coded in two highly symbolic acts, and in a classical academic context no less. Such events speak directly to the classically "loopy" nature of such super natural events—that is, the intentional ways that they interact with their experiencers and make any ideal of "objectivity" look silly. Just how, exactly, would one "prove" any of this with the scientific method?

If I may hazard a guess, I would say that the messages communicated by such events are fairly clear: "You academics will not be able to deal with this" (hence the fainting); and "I told you so" (hence the repeated spilled water glass). And so the anthropologist ends her confession-essay: "My opening anecdote, then, about the ghost of my dead cat overturning a water glass was immediately preceded by an overturned water glass. Make of that story what you will."[18]

I maintain that the academy, including the scientific academy, has *always* been haunted. If we acknowledge this, the usual boundaries between the academic disciplines and our conventional policed ways of knowing will break down, more or less exactly as they did in Fernando's "Supernatureculture" title.

There lies the promise—that is, in new, future ways of knowing that fuse not only the normal and the paranormal but also the natural and the cultural. And it is the "super" that fuses the "nature" of the sciences and the "culture" of the social sciences and the humanities. *There* is where we need to go now, if we are ever to understand that there is no "nature" separate from "culture."

## *The Humanities: The Study of Consciousness Coded in Culture*

Humanists have so far refused to admit this. They have much pre-ferred a truly depressing rewrite of *The Incredible Shrinking Man.* Indeed, the recent half-century history of the humanities could well be the subject of a new movie called *The Incredible Shrinking Human Nature.*

Humanist intellectuals have paid a heavy price for their shrink-ing act. We are more or less ignored now both by the general pub-lic and by our colleagues in the social and natural sciences, whose disciplines make no sense at all outside of universal observations and often work out of bold cosmic visions, wildly counterintuitive models (think ghostlike multiverses and "teleporting" particles) and evolutionary spans of time that make our humanist "histo-ries" look tiny and insignificant by comparison.

I am aware, however, that there are signs of life here. I am thinking in particular of the development of "big history" in his-toriography and of the new materialisms and panpsychisms of contemporary philosophy.[19] These same developments were evi-dent in the recent well-publicized doubts about the adequacy of neo-Darwinian materialism expressed by the philosopher Thomas Nagel in *Mind and Cosmos.*[20] I think these are all positive signs, but I wonder if these are finally bold enough. A new materialism, after all, is still a materialism, and the big histories remain aller-gic to any hint that human beings may be more than historical beings. In short, these new moves still keep the game-changing flips and anomalous material off the table; as a result, their conclu-sions are still exclusions.

I also wonder if we in the humanities are not ignored for good reasons. Why, after all, *should* anyone listen to a set of disciplines whose central arguments often boil down to the claim that the only truth to have is that there is no truth; that all efforts toward truth are nothing more than power grabs; and, finally, that all deep conversation across cultural and temporal boundaries is essentially illusory, that we are all, in effect, locked into our local language games, condemned to watching shadows in our heads, which are going nowhere and mean absolutely nothing. We have lost any sense of the universal, any sense of the human as human.

There may be a way out of our present impasse if we put these kinds of extreme, anomalous, outrageous narratives in the middle of the table. Once we do this, things that were once impossible to imagine will soon become possible to think, theorize, and even test. We may find that we actually *need* these "impossible" things to come up with better answers to our most pressing questions, including the biggest question of all: the nature of consciousness.

I propose that we reimagine the humanities as *the study of consciousness coded in culture.*[21] Not that human reason and our ordinary forms of awareness have immediate access to consciousness as such. I understand that we can only study consciousness indirectly—that is, as it is reflected and refracted in cultural artifacts in the humanities (literature, art, language, religious expression); or in social expressions in the social sciences (institutions, cultural practices, legal systems, voting patterns); or in brain-based neurological processes (cognition, perception, temporality) in psychology and neuroscience.

This little proposed shift would be enough, I suspect, to bring the humanities back to consciousness and humanists back to the debate as central and valued participants. The humanities would no longer be, as my Rice University colleague Timothy Morton teasingly puts it, "candy sprinkles on the cake of science." Quite the contrary, our texts, our narratives, and our methods of interpretation would now function as central clues or pointers to where *anyone* interested in the nature of mind might go for real answers. We would have something to say—something really, really important.

Is this really such an outrageous suggestion? Consciousness is the fundamental ground of all that we know, or ever will know. It is the ground of *all* of the sciences, *all* of the arts, *all* of the social sciences, *all* of the humanities, indeed *all* human knowledge and experience. Moreover, as far as we can tell at the moment, it is entirely sui generis. We know of nothing else like it in the universe, and all we know we know only in, through, and because of this same consciousness. There are those who claim that consciousness is not its own thing, is reducible to warm, wet tissue and brainhood. But to this day no one has come even close to showing how this might work.

Probably because it doesn't.

### *A Very Short History of the Filter or Transmission Thesis*

As scholars like the American literary critic Victoria Nelson in *The Secret Life of Puppets* or the Dutch historian of Western esotericism Wouter Hanegraaff in *Esotericism and the Academy* have demonstrated, Western intellectual history has seen immense swings back and forth between Platonism and Aristotelianism—that is, between a visionary philosophy rooted in mystical and visionary experience (a Platonism which helped produce, among countless other things, the conviction that profound mathematical and philosophical truths are "remembered" or "discovered" and not "constructed") and an empirical rationalism that bases its knowledge on sense data and linear logic. The latter rational empiricism, with the rise of modern science, has been dominant for the last few centuries.[22]

"Every man," Samuel Taylor Coleridge once quipped, "is born an Aristotelian or a Platonist."[23] That may be, but even the Platonists are being educated into Aristotelians these days. Hence Nelson's diagnosis of our present intellectual condition: "The greatest taboo among serious intellectuals of the century just behind us, in fact, proved to be none of the 'transgressions' itemized by postmodern thinkers: it was, rather, the heresy of challenging a materialist worldview."[24]

The solution is not simply to swing back to some kind of pure Platonism, but to effect a synthesis or union of the two modes of human knowing. The sciences are a big help here, and for two reasons. First, because they can challenge humanists to abandon their complete constructivisms and relativisms. And second, because they have utterly failed to explain consciousness.

We now have two models of the brain and its relationship to mind, an Aristotelian one and a Platonic one, both of which fit the neuroscientific data well enough: the reigning production model (mind equals brain, full stop), and the much older but now suppressed transmission or filter model (mind is experienced through or mediated/shaped/reduced/translated by brain but exists in its own right "outside" the skull cavity[25]).

The latter, suppressed model, we might suggest, is more or less equivalent to Plato's "Sun" outside the "cave" of the skull in his famous parable of the cave. The parable involves a group of prisoners chained to the floor of a cave and condemned to watch shadows on the wall in front of them, until one of them escapes and emerges from the darkness into the bright daytime sun. Blinded at first, he eventually adjusts to a vision of the real world. When he returns to his fellow prisoners to tell them the truth of things, they do not believe him. He sounds like a madman. They know what they see. They know what is real. And they cannot be told otherwise.

Pretty much everything hinges on whether we can integrate these two models now—that is, whether we can resist an either/or solution. So far, we have not been able to. The rules of the game we are playing are defined by the complete dominance of the first production model and the near total suppression of the second transmission model, which now sounds "mad" or "crazy." In short, Plato may admire Aristotle, but Aristotle sneers at Plato as a fool. The humanities read and admire the sciences, but the sciences generally completely ignore the humanities, or worse.

Though this is not always the case. Consider the musings of one contemporary neuroscientist, David Eagleman. At the very end of his book *Incognito*, Eagleman turns to the question of the soul and takes on promissory materialism, the commonly heard claim that although we do not yet know how to explain this or that (in this case, "mind") through causal mechanisms and physicalist processes, we eventually will be able to. Indeed, promissory materialism states that someday *everything* will be explained in a materialist framework, because everything is finally only matter. The argument is circular. It simply presumes that which it is trying to establish.

It is also historically ill-informed, as any good history of science can demonstrate. It is extremely unlikely that we just happen to be living at the historical moment when all things will soon be explained within *any* framework, materialist or otherwise. This belief is a species of what is sometimes called "currentism" in the philosophy of science, since it privileges the current state of science as somehow final and infallible. Eagleman observes it is more likely that both "mind" and "matter" will get stranger the more we learn about each, and so also will our models of their assumed relationship. Hence his final parable:

> Imagine that you are a Kalahari Bushman and that you stumble upon a transistor radio in the sand. You might pick it up, twiddle the knobs, and suddenly, to your surprise, hear voices streaming out of this strange little box. . . . Now let's say you begin a careful, scientific study of what causes the voices. You notice that each time you pull out the green wire, the voices stop. When you put the wire back on its contact, the voices begin again, . . . you come to a clear conclusion: the voices depend entirely on the integrity of the circuitry. At some point, a young person asks you how some simple loops of electrical signals can engender music and conversations, and you admit that you don't know—but you insist that your science is about to crack that problem at any moment.

Assuming that you are truly isolated, what you do not know is pretty much everything that you need to know: radio waves, electromagnetism, distant cities, radio stations, and modern civilization—that is, everything "outside" the radio box. You would not even have the capacity to *imagine* such things. And even if you could, "you have no technology to demonstrate the existence of the waves, and everyone justifiably points out that the onus is on you to convince them." You could convince almost no one, and you yourself would probably reject the existence of such mysterious, spiritlike waves. You would become a "radio materialist." "I'm not asserting that the brain is like a radio," Eagleman writes, "but I *am* pointing out that it *could* be true. There is nothing in our current science that rules this out."[26]

Translated into the present discussion, I might gloss Eagleman by observing that there is nothing in the "Aristotelian" or present scientific models of the brain that rules out the "Platonic" models of the mind as primary and prior to the brain. The shadows on the wall of the cave cannot rule out the Sun outside. It is perfectly possible, and perfectly reasonable, to affirm both.

The radio parable may be Eagleman's, but the basic idea and analogy are not new. Plato's parable of the cave is 2,400 years old. And it is Plato's Academy that lies at the origins of the modern university. Something to think about.

If we want something a bit closer to home, we can point to the Harvard psychologist and philosopher William James, who in 1909 had turned to the same "radio" analogy even before there were radios. He explained his transmission theory of mind by comparing it to Marconi's wireless telegraph, a device that could receive and transmit radio waves.

Twain had already written of "mental telegraphy." And James himself had already written and spoken of a similar idea under the banner of a "transmission" theory of mind in his 1897 Ingersoll Lecture on Human Immortality at Harvard. There he had signaled that consciousness is a function of the brain, but that *function* is not the same thing as *production*. Function can also denote transmission, as when a prism refracts a light that is not produced by the prism itself. And so: "When we think of the law that thought is a function of the brain, we are not required to think of productive function only; *we are entitled also to consider permissive or transmissive function*. And this the ordinary psychophysiologist leaves out of his account."[27] James called radio materialists "medical materialists."

A bit later, when there were radios everywhere, the French philosopher and Nobel laureate Henri Bergson would employ the radio analogy to elucidate his own philosophy of mind and his theory of multiple worlds, even if our sensory system has evolved to pick up and interact with only a sliver of these. G. William Barnard calls this Bergson's "radio reception theory of consciousness."[28] In it, our "attunement" to the world is accurate enough to survive and adapt and generally get along, but it by no means exhausts what is there. Here is Bergson:

> Nothing would prevent other worlds corresponding to another choice, from existing with it in the same place and the same time: in this way twenty different broadcasting stations throw out simultaneously twenty different concerts which coexist without any one of them mingling its sounds with the music of another, each one being heard, complete and alone, in the apparatus which has chosen for its reception the wave-length of that particular station.[29]

In such a model, what we experience as "reality" is a function of our evolved cognitive and sensory systems interacting with the real in very selective and limited ways. Put most simply, *our reality is a pragmatic one*, not an exhaustive or even a particularly accurate one.

Plato, James, Bergson, and now Eagleman—all saying different but related things—are only the beginning. There are *countless* clues in the history of religions that rule the radio theory in, and that suggest—though hardly prove—that the human brain may well function as a superevolved neurological radio or television and, at least in rare but revealing moments, when the channel suddenly "switches" (or "flips"), as an imperfect receiver of some transhuman signal.

Although it remains a metaphor and as such is hardly perfect, the beauty of the radio or transmission model is that it is symmetrical, intellectually generous, and—above all—capable of modeling what we actually see in the historical data (when we really look). It affirms everything that we have been doing for the last century or so in the humanities and the sciences (that is, all that Aristotelian stuff about the body and the brain), *and* it puts back into the game much of the evidence that we have ignored as impossible or nonexistent (that is, all that Platonic stuff about the Sun of consciousness).

Such a symmetrical "radio model" has no problem with Mark Twain's knowing about his brother's future funeral weeks ahead of time, the wife's empirically accurate dream about her husband's distant car wreck, the mother's bathtime "seeing" of her daughter lying on some train tracks, or the Swedish scientist-seer's moment-to-moment clairvoyant tracking of a neighborhood fire fifty miles away. The mind can know things distant in space and time because it is not finally limited to space or time. Mind is not "in" the radio. The payoff here is immense: The impossible suddenly becomes possible. Indeed, it becomes predictable. And our range of understanding becomes *immense*. So, too, do our sciences and our humanities (take that either way).

What we have been doing for the last few centuries is studying the construction and workings of the physical radio. But the radio was built for the radio signal (and vice versa). How can we possibly understand the one without the other? It is time to come to terms with both. It is time to invite Plato back to the table. It is time to restore the humanities to consciousness.

But this is all too dualistic, too simple. It is also time to recognize that every metaphor is an inadequate one, and that all such radio (or now computer) metaphors will need to be replaced by others, which we have not yet even imagined, no doubt because we have not yet created the technologies or discovered the sciences that will produce these new ways of imagining the world and ourselves. These future metaphors, of course, will also be inadequate. The super natural world will remain just that—super to us.

# 2
## FLIPPED SCIENTISTS

*As an M.D. with a long career at esteemed medical institutions like Duke and Harvard, I was the perfect understanding skeptic. I was the guy who, if you told me about your NDE [near-death experience], or the visit you'd received from your dead aunt to tell you that all was well with her, would have looked at you and said, sympathetically but definitively, that it was a fantasy.*

—EBEN ALEXANDER, M.D., *Map of Heaven*

One could at this point begin listing dozens, hundreds, and then thousands of similar reports of the unexplained and the extraordinary—the many-headed Hydra of the history of anomalous experience. I suspect that most intellectuals, scientists, and medical professionals (part of my respected audience here) would reflexively cut off the vast majority of these heads. The assumption would be that the reporter must have been naïve, or simply did not know better, since such nonsense obviously conflicts with science.

## *The Sociology of Scientific Nonsense*

Science is a method, however, not a set of dogmas with which one can conflict, and it has progressed so spectacularly over the last few centuries by demonstrating, over and over again, that what was previously thought to be nonsense is, in fact, almost certainly the case. Science makes the impossible possible.

Have you ever actually sat down and listened to what quantum physicists or cosmologists are saying about the nature of matter and the structure of the universe? To any ordinary mortal, it is all pure craziness: matter as congealed energy, the relativity of space and time, nonlocality, entangled particles that communicate instantly across space-time, multiple dimensions that the primate brain simply cannot imagine or picture, dark energy that, we are now told, constitutes the vast majority of the universe, a "big bang" out of nowhere and nothing, parallel universes, and on and on we go into one impossible thing after another. And yet, we are told, this is precisely what the empirical evidence and mathematics point toward. Indeed, we are repeatedly told by the physicists themselves that if quantum mechanics makes sense to you, you clearly do not understand what is being said. Literal nonsense (as in "You cannot understand this with your sense-based imagination and reasons") is the very mark of the quantum truth.

Science is all about exploring anomalies and taking them up as the keys of future scientific knowledge, which is often nonsense to our common sense. The apparent "nonsensical" nature of extreme religious experiences, moreover, often fits *seamlessly* into the "nonsensical" models of contemporary science. I am thinking, for example, of the impossible 360-degree vision often reported in out-of-body and near-death experiences—exactly what one would expect if a person had suddenly popped into an extra space-time dimension. The new science, here the multidimensional geometry of space-time, actually renders plausible and possible what was previously thought to be implausible and impossible, even as the anomalous experiences throw a new light, a human light, on the multidimensional geometry.

A scientist might hesitate to report anomalous experiences because of his or her social and professional context and the sociology of this particular mode of knowledge. There are subtle and explicit forms of censorship that effectively suppress such reports to protect the present reigning interpretation of the world—classical or conventional materialism. We do not call these "disciplines" for nothing. They discipline us, and they discipline the world.

A poignant and powerful example of this professional censorship is the story that the social worker Kimberly Clark Sharp told her colleague, the psychiatrist Bruce Greyson of the University of Virginia School of Medicine. This story, which comes from the very heart of modern professional medicine, revolves around a male scientist who had a near-death experience after a heart attack. Dr. Greyson explains why it was so difficult for the man to speak to his wife about his unbelievable experience, and what happened when he did:

> He and his wife were both hard scientists in the medical school. He had an NDE with his heart attack, and he asked Kim to help him talk to his wife about it. The three of them are sitting in the room. The man is telling the story, and he can't bear to look at his wife because he *knows* what she is going to think.

She is going to think that what he is saying is scientific nonsense.

When Kim finally looks over at the man's wife,
her mouth has dropped open.

She says, "Stop a minute. Stop a minute."

[Kim] says to the man's wife, "What is going on
with you?"

She says, "This happened to me ten years ago. I
was afraid to tell anyone about it."[1]

And so it goes. The materialist interpretation of the world and of
science itself is protected not by the facts or by the data of our honest experiences, but by what is essentially social and professional
peer pressure, something more akin to the grade-school playground
or high school prom. The world is preserved through eyes rolling
back, snide remarks, arrogant smirks and subtle, or not so subtle,
social cues, and a kind of professional (or conjugal) shaming.

The medical school couple are hardly alone in terms of their
kept secrets. The history of science and medicine is chock-full of
such incidents. Scientists and medical professionals are human
beings, too, after all, and, like other human beings, they often
have extraordinary experiences that point them beyond their
present relative worldviews to something much more grand and
much more interesting.[2] Such historical events demonstrate in
bright colors that having such an experience need not have anything to do with some form of scientific ignorance, and that
sometimes these same rogue experiences lie at the secret origins
of new knowledge and new science.

One of my basic convictions is that scientists and medical
professionals are now sitting at the very center of the production
of new mystical and visionary literature, and that this same literature signals the early beginning of a new worldview or a new
real. Indeed, it is *precisely* because of their scientific and medical

training that these professionals make such convincing visionaries and authors. This is also why their stories are so compelling.

At the end of the day, what we have here are modern conversion stories, conversions not to this or that religion, but to a new cosmic outlook in which mind or consciousness is primary. The word *conversion*, by the way, comes from the Latin for "turn around"—a specifically religious form of what I am calling the flip.

A good conversion story tells us as much about the previous worldview that is left behind as it does about the new one that is being embraced and celebrated. A conversion story works only if there is a dramatic movement from one world to another. Our modern scientific mystics are so important because they can speak authoritatively to the conventional materialistic worldview that directs how most of us live, think, and feel. They can also tell us why this worldview is limited and incomplete. They know perfectly well how and why their experiences conflict with their earlier outlooks, and they say so.

### Hans Berger

Hans Berger (1873–1941) was the first human being to make an EEG, or electroencephalogram, recording of the human brain, which he did in the 1920s. He also gave the EEG its name.[3] After beginning studies in astronomy and mathematics at the University of Berlin, Berger took a year off, in 1892, and joined the German military. One spring day, during a training exercise, he was tossed from his horse into the direct path of a speeding carriage carrying an artillery gun. He was about to be killed, and he knew it, but the driver of the carriage somehow managed to stop in time.

Just then, Hans's older sister, many miles away, was overcome with great fear and deep dread. She was certain something terrible had happened to her beloved brother. She was so certain that she made her father send Hans a telegram immediately. When Hans read the telegram later that evening, he became convinced that somehow his sister had known at a significant distance about his near-fatal accident:

> He had never before received a telegram from his family, and Berger struggled to understand this incredible coincidence based on the principles of natural science. There seemed to be no escaping the conclusion that Berger's intense feelings of terror had assumed a physical form and reached his sister several hundreds of miles away—in other words, Berger and his sister had communicated by mental telepathy. Berger never forgot this experience, and it marked the starting point of a life-long career in psychophysics.[4]

*Psychophysics.* The strange word already encoded the mind-matter problem—that is, how mental phenomena are, or are not, connected to physical laws. Hans had just experienced, firsthand, a psychophysical event of incredible strangeness and significance that strongly suggested that things were not quite what they seemed. The event seemed to signal that human emotion and thought are not restricted to the skull and brain.

After his military service was over, Berger returned to his university studies, this time to study medicine and become a psychiatrist so that he could pursue this mysterious "psychic energy," as he called it, an energy that could somehow transcend local space and link one brain to another. Eventually, he did, in fact, discover a technological means to record brain waves and demonstrate that the brain was an electrical organ whose activity could be correlated with specific states of mind and mood. As he accomplished all of this, he also engaged in a systematic study of two hundred individuals, each of whom he tested for telepathic abilities after they were put in a trance state. He was obsessed, in the words of his biographer, with "the correlation between objective activity in the brain and subjective psychic phenomena."[5]

It was Hans Berger who coined the term *brain waves*, now omnipresent in our culture and language. Originally, these were called "Berger rhythms" or "Berger waves." Now they are referred to more technically as alpha waves and are associated with the EEG, which depends on Berger's earlier work and invention.

## A. J. Ayer

Prominent British atheist philosopher and logical positivist A. J. Ayer (1910–1989) was suffering from pneumonia and recovering in a hospital bed when he tried to gobble down a piece of smoked salmon that his friends had sneaked him. It went down the wrong way, and his heartbeat plummeted and then stopped for four full minutes. The medical staff managed to bring him back. The first thing he said was, "You are all mad."

Ayer himself confessed that he did not know how to interpret his own strange hospital-bed oracle. He offered two completely different interpretations. In the first interpretation, he suggested that the phrase was addressed to his Christian colleagues: They were all mad for believing in a heaven, for he found none on the other side. The second interpretation he offered was that the phrase was addressed to his skeptical colleagues, for he seemed to have found "something" on the other side. He tended toward the former reading, but he admitted either was possible. Let me observe here that this kind of ambiguity is entirely typical of paranormal experiences: They are often structurally and irreducibly paradoxical, and their ambiguity cannot be finally resolved. And *that* seems to be a big part of their punch and point.

Note Ayer's use of the vague but honest word *something*. We have already encountered it in the story of the Danish divinity student Johanne and her dark vision of the burning corpses of the theology professors in the sky. Behind all of this, she caught, in her own words, "a glimpse of something." The "something" that Johanne glimpsed turned all the Bible into "bullshit." The "something" the philosopher found on the door of death was not quite so irreverent, but it was equally, bizarrely visionary, like Johanne's burning, floating corpses. Ayer found himself trying to cross a river, twice, in his near-death experience: a visionary fact, he pointed out, that was no doubt indebted to his classical training and his early exposure to Greek mythology's river Styx.

But there were weirder things to note. His thoughts became persons. He was confronted with a "red light" that he knew was "responsible for the government of the universe." He sensed in the experience that space was somehow out of joint, and that it was his job to fix it. Knowing his Einsteinian physics (apparently, Einstein is relevant even after death), he decided that if he could fix time, he could also fix space, since space-time can be treated as a single unit. He was also informed of "creatures" or "ministers" who had been put in charge of space and time. It seems they were not quite doing their jobs.

Ayer did not take this literally, although he seemed inclined to think that there might be something "empirical" or "objectively real" about the red light, as another near-death experience he had been told about contained the same phenomenological feature.[6] His general philosophical conclusion was this: "On the face of it, these experiences, on the assumption that the last one [the other reported near-death experience] was veridical, are rather strong evidence that *death does not put an end to consciousness.*"

Ayer recognized that his brain may well have been active,

despite the fact his heart had stopped pumping blood to it. He also insisted on the fallacy of assuming that "a proof of an after-life would also be a proof of the existence of a deity." There is an important point, which is seldom pointed out: Again, the paranormal need not be interpreted religiously, much less theistically. Ayer himself remained a steadfast atheist. But he also saw that experiences like his own might well help clarify in the future the philosophical question of the "*relation between mind and body.*" His own assessment was that his near-death experience has "slightly weakened" his conviction that his death "will be the end of me, though I continue to hope that it will be."[7]

In a "Postscript to a Postmortem," written a month and a half later, Ayer sharpened his cut as he tried to correct people's mistaken readings of his original piece. He wished to emphasize that he thought the most likely explanation of his own experience was that his brain was still functioning and so producing the visions (although how that works, no one can quite say). He also wanted to stress that he had been careless in his writing, and that his own personal convictions in not surviving death had not weakened. What *had* weakened was his polemical desire to deny the logical plausibility of an afterlife and his "inflexible attitude towards that belief."

He then went on to suggest that there may be "experiences which do not belong to anybody; experiences which exist on their own." He then launched into a logical discussion of the philosophical problems with the Christian doctrine of the resurrection of the body, and how reincarnation presents fewer difficulties for the thinker (even though, he pointed out, he had earlier ridiculed the same belief). "But our concepts are not sacrosanct. They can be modified if they cease to be well adapted to our experience."

Ayer pointed out that all we would need to seriously consider (which is not to say accept or believe) reincarnation is good empirical evidence that enough individuals could remember in sufficient detail their previous lives, and that these memories could be verified. Ayer may not have known it, but we have exactly this evidence, and in abundance, in the extensive global archival base on childhood reincarnation memories and strange birthmarks collected by the psychiatrist Ian Stevenson at the University of Virginia, which now stand at about 2,700 cases. But that is another story, which has been told and documented numerous times and which noted skeptics have described as impressive.[8]

Ayer saw clearly that no belief from the history of religions or from modern science fiction can quite compete with the outrageous claims of the physicists and cosmologists. Just go read Stephen Hawking's *A Brief History of Time*, Ayer suggested. The book is a best-seller, but "[p]erhaps the reading public has not clearly understood what his speculations imply. We are told, for example, that there may be a reversal in the direction of the arrow of time. This would provide for much stranger possibilities than that of a rebirth following one's death. It would entail that in any given person's life a person's death preceded his birth. That would indeed be a shock to common sense."[9] Ayer had a sense of humor up to the very end.

The point of all of this for us is not that reincarnation is true, or that there is an afterlife, or that we survive our physical deaths, or that time's arrow can be reversed. The point is that A. J. Ayer, one of the most rigorous philosophical thinkers of the twentieth century, was willing and able to think entirely outside his familiar boxes about these subjects after his near-death experience, *but not before it.* He was willing to be flipped, even if he never quite flipped, not at least in this life that those of us who have not died can perceive and confirm. His famous logic was consistent and honest until the very end.

### Eben Alexander

In 2012, neurosurgeon Eben Alexander III published his *Proof of Heaven: A Neurosurgeon's Journey into the Afterlife*.[10] In 2014, he followed up his best-seller with *The Map of Heaven: How Science, Religion, and Ordinary People Are Proving the Afterlife*. In 2017, he published a third book, this one with coauthor Karen Newell, *Living in a Mindful Universe: A Neurosurgeon's Journey into the Heart of Consciousness*. Here we have a trilogy from an accomplished brain surgeon from the very heart of the medical profession entering the public imagination after his own extreme spiritual experience to shift, really reverse, assumptions about the mind and its relationship to the brain.

The three books, moreover, give ample witness of a scientifically trained mind, inspired and empowered but also deeply shaken by an extreme anomalous event, seeking out and learning from the riches of humanistic, philosophical, and mystical literature. It is not a scientifically trained mind telling historians and humanists the truth of their own literature (and their own brains), but a refreshingly reciprocal two-way conversation.

Alexander completed his M.D. at Duke University School of Medicine and spent fifteen years on the faculty of Harvard Medical School. He shared the usual medical reductive convictions about the material nature of mind: Mind is brain, and nothing more. That was before November 10, 2008, when, at the age of fifty-four, Alexander began experiencing intense pain in his back and head and collapsed into a deep coma for seven days. The coma, it turns out, was brought on by a rare case of E. coli meningitis. Because of the severity of the infection, its weeklong length, and its damage, his attending doctors gave him a 2 percent chance of recovery, and a zero percent chance of fully recovering his faculties and having any quality of life.[11]

While in the coma, Alexander encountered, in his own emphasized words, "a world of consciousness that existed *completely free of the limitations of my physical brain.*" This world may or may not have been independent from his brain, but it was certainly formed around some classical religious themes—particularly spiritual flight, altered states of energy and light, heavenly music, and a profound sense of unity and love. This was also a world organized around some vivid visionary stages, to which he gives capitalized names: the Realm of the Earthworm's Eye View, the Spinning Melody, the Gateway, the Girl on the Butterfly Wing (who is also encountered as an Orb of Light), and, finally, the Core.

Such capitalized themes suggest strongly that the narrative had been polished and reworked from whatever original memories Alexander recalled, but this is entirely typical of extreme religious experiences and is precisely why we need the tools of the humanities (which revolve around the interpretation of such narratives and redacted texts) to really understand what is going on here, from the original private subjective experiences to the published text and its public reception and various competing readings.

It is a distinct feature of modern mystical literature that extraordinary experiences like those of Alexander often take place or are immediately interpreted within an evolutionary framework deeply indebted to Darwin's revolution but nevertheless non-Darwinian (that is, teleological or goal-oriented) in structure and intent. I have referred to these modern narratives elsewhere as forms of evolutionary esotericism—that is, as modern spiritualities that draw on the models of modern science but cannot be fitted into either traditional religious or conventional scientific paradigms.[12] Alexander's near-death experience (NDE) and later interpretations fit neatly into this pattern.

While in the Realm of the Earthworm's Eye View, for example, he describes how he was in a "primordial" state. It was "as if I had regressed back to some state of being from the very beginnings of life." He felt like a mole or a worm. "I wasn't human while I was in this place. I wasn't even animal. I was something before, and below, all that."[13] Later, in *The Map of Heaven*, he would compare what he knew here to the dark, shadowy conceptions of death of ancient Greece, before the mystery religions brightened the afterlife with initiatory experiences and secret teachings of immortality. One might also invoke the ancient Hebrew notion of sheol here—a place of darkness and death that all were believed to go to when they departed this bright life.

It was the Spinning Melody that rescued him from this darkness, from this muddy, earthy world—a beautiful white musical light that spun slowly, casting out living golden and silver filaments, as it called to him through what he would later call a "rip" or a "portal."[14] It was in this way that the Spinning Melody acted as the means through which he entered what he calls the Gateway Valley, a paradisiacal natural landscape that he experienced as "completely real" or "hyper-real" and compares to Plato's world of forms.[15] This world featured millions of butterflies, gorgeous valleys, streams, waterfalls, people, dogs, and, high above the clouds, "flocks of transparent orbs" or "shimmering beings" that resembled birds or angels but which were "more advanced. *Higher.*"[16] He also describes "a vast array of larger universes that took the form of what I came to call the 'over-sphere,' that was there to help in imparting the lessons I was to learn."[17] He was riding on one of those butterflies, "as a speck of awareness," accompanied by a girl in a powder blue and indigo dress whom he did not recognize but whom he later learned from a photograph was his biological sister, Betsy.

An empirical "hit" drops into the visionary landscape. He had never met Betsy before she died and so did not know what she looked like (since he was adopted at birth). The photograph of Betsy, in which he clearly recognized the young woman of his earlier NDE, came as a severe shock, a shock of recognition that the near-death experience was no hallucination. He now knew with a new certainty: The afterlife was *real*.[18]

Alexander communicated with the beings he met in that other world directly, without language. This is an extremely common trope within anomalous events, a feature for which most experiencers use the traditional word *telepathy*. We have already encountered the word briefly, in Hans Berger's experience with his sister. It had already been in circulation in 1892 for about a decade, when Hans was thrown from his horse. The noun was coined in 1882 by the Cambridge classicist Frederic Myers (after studying and analyzing thousands of case studies) to name all of those highly charged traumatic or "emotional" (*pathos*) messages that are communicated from an otherwise-unexplainable "distance" (*tele-*). No playing cards or stock market predictions here: This was all about death, suffering, and love—that is, profoundly human experiences that no scientific experiment can predict, control, or measure.

Significantly, Alexander also notes that he understood things directly and instantly in the other world, things that would have taken him years "here," in the ordinary world, to process and understand. In his classic *The Varieties of Religious Experience*, William James called these moments "noetic." What he meant by this Greek-based word is that such experiences carry a direct knowledge that cannot be reduced or explained by any simple cognition or sensory input. One can think of this as knowing something directly without or outside the "filter" of the brain and its various cognitive and sensory mediations. Consciousness, freed from those buffers and blocks, simply *knows*. Here is another future form of knowledge that is also ancient.

Alexander then entered "an immense void, completely dark, infinite in size, yet also infinitely comforting." This was what he calls the Core. The Core was God as "mother" or "giant cosmic womb." There he learned that there are "countless higher dimensions" from which one can enter our world at any place or time.[19] Most of all, he learned that love is "the basis of everything."[20] In his third book, after he had encountered comparative mystical literature in some depth, he would describe the Core as "the source of *all*, the ultimate noduality of pure oneness."[21]

Alexander explicitly and repeatedly invokes the filter thesis and the neuroanatomy of the bilateral brain to explain why we normally cannot access such truths. Indeed, he explicitly cites the most extensive contemporary work on the filter thesis, *Irreducible Mind* (2007), by the Harvard-trained neuroscientist Edward Kelly and his colleagues.[22] He also expresses amazement that Dr. Bruce Greyson, the doyen of near-death studies, and Dr. Kelly were colleagues and lived and worked a mere ninety minutes from his door, at the University of Virginia.[23] The model was always there, and it has long sent deep taproots into the academy and the medical profession. He simply was unaware of it.

In his own expression of the filter thesis, Alexander writes that our "brain blocks out, or veils, that larger cosmic background."[24] In an interview with NPR host Steve Paulson, Alexander expanded on this idea. There he explained how we are being "dumbed down" by our bodies. He spoke of the "shackles" of our brain and body and how "we are conscious in spite of our brain," not because of it. All of this is necessary, he speculated, since if we were in constant touch with that other world, we would not be able to survive as mammals in this one. He even referred to our individual forms of mind as a "divine spark" and the brain as a kind "reducing valve" actively reducing our access to and awareness of this spark of divinity.

Well within the evolutionary esoteric mode, Alexander believes that he was given a "foretaste" of our evolutionary future, of our far-future form of knowledge. He has little practical hope in explaining all of this to us here and now, though. His experience, after all, was "rather like being a chimpanzee, becoming human for a single day to experience all of the wonders of human knowledge, and then returning to one's chimp friends and trying to tell them what it was like knowing several different Romance languages, the calculus, and the immense scale of the universe."[25] This is a clear return to Plato's parable of the cave, with a little Darwin added to the mix.

Such evolutionary esoteric implications are drawn out further in the third book, where he links the filter thesis and the idea of human potential, a particularly potent science-spirit fusion drawn from a whole host of previous nineteenth- and twentieth-century writers of whom he is now aware: Frederic Myers, William James, Henri Bergson, F. C. S. Schiller, and Aldous Huxley. Alexander invokes again the filter thesis, the hypothesis that "conscious awareness can be liberated to a much higher level when freed up from the shackles of the physical brain, as happened while I was in coma." He then goes on:

> The scientific implications are stunning, and provide
> powerfully for the reality of the afterlife. But this is
> only the beginning. As we come to realize that exam-
> ples of exceptional human potential (as in genius-level

creativity, telepathy, psychokinesis, precognition, and past-life memories) really occur in some people, we begin to realize that the latent ability is there in all humans. In other words, these are skills that one can cultivate and enhance.[26]

Later in the same book, he recounts numerous likely instances of psychokinesis in his own post-NDE life involving a "crackling," electriclike sense of energy and light around his own body, streetlights going out while he walked below them (a very common phenomenon with its own literature, by the way), an inability to keep electronic devices (a computer and three watches) running properly in his presence, and, finally and most dramatically, a tree that crashed and fell, twice, while he manically wrote his first book, *Proof of Heaven*, a few feet away in his home.[27] He also writes of something that he calls "deep time," which he bases on his experience in the coma when the flow of time "was wrapped into a tight loop, or even a point." In such deep time, time itself can seem to move forward or backward, or our awareness "can jump to regions remotely distant from a consensus 'earth time.'"[28]

As extreme and extraordinary as this all sounds, *all* of these phenomena are very common and well known in the literature on anomalous experience.

Alexander's final takeaway? That "conscious awareness can exist entirely independent of the brain," and that only a fraction of what consciousness is can be captured in language and communicated to those who have not "been there." The extraordinary experience, in other words, is necessary. One cannot think or reason or experiment one's way there. Accordingly, Alexander now knows many things that his peers and medical colleagues, who have not "been there," do not know. Foremost among these things is that the materialist model upon which conventional science is based is "fundamentally flawed." "At its core," he writes, this materialism "intentionally ignores what I believe is the fundament of all existence — the nature of consciousness."[29] Not that religion has it right, either. Hence Alexander's flip and the third way of thinking it inspired in him:

> I'm a doctor who had an NDE—a solid member of the "dogmatic science" side of the room, who had an experience that sent him over to the other side. Not the "dogmatic religion" side, but a third side of the room, if you will: a side that believes science and religion both have things to teach us, but that neither has, or ever will have all the answers. This side of the room believes that we are on the edge of something genuinely new: a marriage of spirituality and science that will change the way we understand and experience ourselves forever.[30]

In his most recent book, he writes that "as we come to realize the primacy of consciousness and adopt the power inherent in full-blown metaphysical idealism, many perceived limitations will disappear."[31] In my own language: As the rules of the game change and new orders of knowledge appear, the impossible will become possible (since it was never really impossible to begin with).

### Barbara Ehrenreich

In early May 1958, a seventeen-year-old teenager named Barbara Ehrenreich went skiing on Mammoth Mountain in Northern California.[32] On the way home, something would happen to her on the streets of a little town called Lone Pine that would eventually upend her entire philosophical worldview. *Living with a Wild*

*God* is her profound and hilarious look back on her science training and this life-changing "epiphany," as she insists on calling it.

Ehrenreich grew up in a passionately atheistic family, signaled
by such stories as the time her dying grandmother threw a crucifix
across the room when a priest tried to give her the last rites. Trained
in these family ways of fighting against all forms of authoritarian-
ism, control, and oppression, Ehrenreich would spend her professional life as a science journalist, a take-no-prisoners writer, and a
social activist picketing and meeting for various feminist and economic justice causes. Lone Pine continued to haunt her, though,
even if she effectively suppressed it. She had known "an event so
strange, so cataclysmic, that I never in all the intervening years
wrote or spoke about it."

> So what do you do with something like this—an
> experience so anomalous, so disconnected from the
> normal life you share with other people, that you
> can't even figure out how to talk about it? I was also,
> I have to admit, afraid of sounding crazy. Try insert-
> ing an account of a mystical experience into a conver-
> sation and you'll likely get the same response as you
> would if you confided that you had been the victim
> of an alien abduction.[33]

The reasons are simple enough. These sorts of experiences, of
which there are countless forms in the history of religions, reveal
to their subjects that the material world is fiercely alive, perhaps
even cosmically conscious. Many of them also suggest that our
world is populated with innumerable strange creatures, which
are normally completely invisible to our perceptual systems.
Conventional science, on the other hand, insists that the physical
world consists only of what Ehrenreich describes as "tiny dead

things, mindless particles following their destinies," bouncing around in orderly ways to produce, eventually, big dead things that think they are conscious agents but are really not.[34]

Ehrenreich, who has a Ph.D. in cell biology and so knows her science as well as anyone, rips into this scientific materialism. She explains how the laboratory wants to work only with dead things and kills living things in order to experiment on them. She herself "sacrificed" twelve hundred mice as a young scientist. She even takes a humorous swipe at one of the pillars of modern physics, the hallowed second law of thermodynamics, which states, in her own gloss, that in a closed system "everything tends toward death, or toward rot, or at least toward extreme boringness."[35] And then she goes further. "The whole project of science," she explains, "is to crush any notion of powerful nonhuman Others," and the vocation of the scientist is "to keep all that was uncanny or unspeakable stuffed out of sight."[36] What to do, then, with her own perfectly empirical experiences of the precise opposite truth—namely, that everything is fiercely *alive*?

There were early signs of what was to come. As an adolescent, Ehrenreich had had a long series of anomalous experiences during which all of her ordinary perceptual ordering of the world—that is, all of her language, ego, and cultural labels—simply dissolved before reality *as it is*. Basically, she fell into what she calls a "total perception" of the real beyond all culture and individuality. She invokes the pathologizing language of "dissociation" to describe these anomalous perceptions beyond perception and ego, but she knows that such talk (which is really a kind of condescending dismissal) seriously misses and distorts what she knew. Occultism and science fiction come much closer. She thus writes of these multiple events as a "breach in the dividing membrane" of reality, as "fissures in reality," as "some alternate realm of being and knowing," and as a "rent in the fabric of space-time."[37]

Then came Lone Pine. In a state of sleep deprivation (she had slept the night before in the car) and hypoglycemia (she had not eaten enough)—that is, "in the kind of condition that the Plains Indians sought in their vision quests"[38]—she woke up and took a walk through town. Or tried to do so. The world burst into living fire. She invokes the biblical "burning bush" to describe her sacred sense of things, but there was little biblical, much less monotheistic, about the experience. Indeed, Ehrenreich makes it crystal clear throughout the book that she despises monotheism as a form of "deicide" that insists on killing every other god and leads eventually to modern science, which kills everything else.

But nothing was dead on the streets of Lone Pine on that fateful morning. *Everything* was spirited. *Everything* was alive, even the teacups and toasters in the shop window of a second-hand store:

> The world flamed into life, . . . This was not the passive beatific merger with "the All," as promised by the Eastern mystics. It was a furious encounter with a living substance that was coming at me through all things at once, . . . Nothing could contain it. Everywhere, "inside" and out, the only condition was overflow.

Once again, she returns to her inability to describe what she knew, what she had become in the fire that had consumed her. What could she have said? "That I had been savaged by a flock of invisible angels—lifted up in a glorious flutter of iridescent feathers, then mauled, emptied of all intent and purpose, and pretty much left for dead?"[39]

All of *Living with a Wild God* is an attempt to contextual-
ize and describe this flock of mauling angels. She moves through
a number of possibilities, but one of her most interesting moves
comes at the very end of the book, when she invokes the Valis
experience of the science fiction writer Philip K. Dick in order to
suggest that what she experienced was a kind of biological god.
"Valis" or "VALIS" was Dick's creative acronym. It stood for Vast
Active Living Intelligence System. Basically, Valis was a cosmic
form of mind that overwhelmed the science fiction author for two
months, in the winter of 1974, and sent him back to his entire
corpus to look for clues that led up to *this*. Dick would spend his
remaining eight years writing furiously in his private journals, and
three final novels, as he tried to fathom the paradoxes (or flips)
of Valis. He never did. But Dick, Ehrenreich points out, sensed
that Valis was not some abstract monotheistic deity, much less a
subjective hallucination, but a *living creature*. His own metaphysical opening was experienced as what he himself described as an
"interspecies symbiosis."[40]

Out of the remembered depths of her own fiery encounter,
Ehrenreich entertains the same notion now, wondering out loud
to what extent that what mauled her that May was a biological species of some kind. She even asks out loud if such invisible beings
"feed" on our forms of consciousness, a *very* common thought, by
the way, in the alien-abduction literature, where it is extended to
our emotions and the bioenergies of our sexual arousals. It is science fiction again that inspires her:

> Science fiction, like religious mythology, can only be
> a stimulant to the imagination, but it is worth consid-
> ering the suggestion it offers, which is the possibility
> of a being (or beings) that in some sense "feeds" off
> of human consciousness, a being no more visible to

us than microbes were to Aristotle, that roams the universe seeking minds open enough for it to enter or otherwise contact.[41]

Ehrenreich appears to be open enough. She confesses that, thanks to her reading of history and theology, she is now ready "to acknowledge the possible existence of conscious beings—'gods,' spirits, extraterrestrials—that normally elude our senses, making themselves known to us only on their own whims and schedules, in the service of their own agendas."[42] She had made the flip through experience, memory, and her reading of science fiction and—at some point, it appears—a bit of the study of religion.

Still, in the end, Ehrenreich's "wild God" is never defined, never really described, never quite captured (hence the "wild" part), but this especially brave author leaves the reader thinking that this deity is multiple, that it is a species, and that it is *after us*. She ends on a haunting, almost predatory note: "I have the impression, growing out of the experiences chronicled here, that it may be seeking us out."[43]

### Marjorie Hines Woollacott

Marjorie Hines Woollacott is a retired neuroscientist. She was teaching at Virginia Polytechnic Institute in 1975. She did not want to go to the meditation retreat, a visit her sister had given her for her birthday. As a professional scientist, Marjorie shared little of her sister's enthusiasm for such things. Still, Marjorie and Cathie were very close. And there was something else. Cathie had given Marjorie a mantra (a single Sanskrit syllable used to focus the mind in many forms of Hindu and Buddhist meditation) to repeat before a plane flight to reduce her anxieties about flying. It somehow oddly calmed her down in ways that went well beyond what she was expecting. Still rather astonished by the odd effects of the mantra on the plane, Marjorie decided to give the retreat a try.

It was a fateful decision.

What the retreat promised was an "initiation" (*diksha*) at the touch of a realized master or guru (guru is Sanskrit for "spiritual teacher"). And these people were not trucking in metaphors. They meant it. The details are well known to historians and ethnographers of modern Hinduism. This is the "descent of power" (*shakti-pat*), an often literally "electric" transmission of subtle energy that is believed to activate a secret physiology or esoteric anatomy in the human body variously described and imagined in different systems. The modern chakra systems are probably the best known of these subtle body systems (*chakra* is Sanskrit for "wheel" and refers here to wheels or centers of energy in the human body located along the spinal channel, from the genitals to the brain core and, in some systems, just "above" the brain[44]).

Subtle body or chakra systems aside, the message is clear enough: The secret of enlightenment sleeps in us all, and it can be awakened by the electric touch of one in whose body it is already awake and active. Hence the importance, even necessity, of the realized saint or awakened guru. The lamp won't go on, after all, if you don't plug it in to the wall socket. One needs a power source. Woollacott uses a similar image: She describes awakening as an unlit candle being set aflame by a lit candle.[45]

The particular guru whom Woollacott encountered on the retreat used peacock feathers and his fingers to help him plug people into the universal current, to light the unlit candle that was Marjorie Woollacott. Here is what happened to Marjorie that day when he touched her:

Then, firmly, I felt the swami's thumb and fingers right between my eyes and on the bridge of my nose. I was alert. My eyes were closed, but my senses were otherwise fully engaged, so that when in this moment I experienced a current of what felt like electricity enter from the swami's fingers into my body, I had a sense of utter certainty about the event. It isn't that I knew precisely what had happened. To this day, I can't explain it. But it seemed as if a tiny lightning bolt leapt from his fingers to a point between my eyes and down to the center of my chest. I could feel the exact point where the energy stopped. I knew it was my heart, not the physical heart but parallel to the physical heart and more like a heart than my physical heart had ever been. I say that because for the first time I could feel energy pulsating from my "new" heart, which seemed to be at my very core, . . . It felt like nectar; it felt like pure love pouring through me. Words went through my mind, and they had nothing to do with scientific analysis: *I'm home, I'm home! My heart is my home!*[46]

The eventual effect of this transmission of energies on Marjorie was profound, life-changing. She would continue her career in neuroscience, but she would also come to see that she has serious doubts about this particular community's reductive, materialist, bottom-up approach to the nature of consciousness. Consciousness, she came to see, is most likely not an emergent property of brain matter, contrary to what everyone around her in her professional life seemed to assume. This only appears to be the case because the scientific method restricts the inquiry to objects or processes that can be measured—that is, to a third-person perspective. Once one shifts to a first-person experience, the inquiry changes, dramatically.[47]

In 2015, Woollacott came out of her professional closet and published *Infinite Awareness: The Awakening of a Scientific Mind*.

After her awakening into the deeper levels of consciousness and energy (and, no doubt, *because* of her awakening), Woollacott sought out professional training in the humanities. She had realized in a very dramatic and personal way that her neuroscientific training was simply not adequate to the task of modeling what had happened to her, much less to explaining it. More specifically, she pursued Asian Studies and the history of a particular philosophical school within Indian religions known as Kashmir Shaivism, so called because of its focus on the Hindu god Shiva and its geographical locus of present-day Kashmir.

This is an especially sophisticated philosophical and mystical tradition well known to scholars of Hinduism. Its basic teaching is that *everything* is consciousness. Everything is literally made of consciousness (*chit*), which manifests in the physical world as a kind of subtle "vibration" (*spanda*), a vibratory energy that congeals or crystallizes into what we perceive and know as material reality. All of this energetic consciousness that takes shape as matter and the physical world is ultimately an efflux, radiation, or emanation of an ultimate Subjectivity, known as Shiva in this Indian tradition or, in Western language, as God. In the vision of Kashmir Shaivism, we might say that everything is an emanation of God, and yet God is more than this everything. Woollacott turns to a Western theological term for this particular teaching. She calls it, quite correctly, *panenetheism*, literally, "all-in-God-ism."

Panentheism is not the traditional theism of the Western monotheisms, in which God "creates" the world and so is separate from it. Nor is this pantheism per se, since this ultimate Subjectivity exists prior to the physical universe that it manifests or radiates. In a classical pantheism, there is no difference between God and world. They are the same thing, and there is nothing left over or beyond this same substance. Not so in panentheism. God and the world are, and are not, the same thing.

Actually, one of the most clever and succinct expressions of panentheism I have ever encountered occurs in the writings of Barbara Ehrenreich, whether or not she is cognizant of the term. As Ehrenreich struggles to express the paradoxical nature of her mystical experiences, at once subjective and objective, she writes that she came to realize that she was both "a part of" and "apart from" the universe. This is a nearly perfect expression of pan-entheism, with the human subject, of course, in the position of God. This, too, by the way, is quite faithful to a tradition like Kashmir Shaivism, where Consciousness as cosmic, as "divine," can be manifested and known through the human teacher. Hence, Marjorie getting zapped by the visiting guru or awakened teacher.

The both/and structure of panentheism can be expressed as a feature of divinity or humanity equally well. It works in either case. It also provides a very distinct answer to the mind-matter interface question. The answer in this system is that there is no ultimate division or separation between mind and matter. Mind is prior and cosmic, and matter is an emanation from that mind that, in turn, shares its nature and "divinity." Mind does not "create" the physical world as something other than itself. Mind "emanates" or "radiates" the physical world as an expression of itself. Matter is congealed mind.

Such a theological system might sound fantastic, but we should consider Woollacott's situation here. How, exactly, would you make sense of getting touched on the forehead by a famous Indian spiritual teacher, after which an apparent little lightning bolt of knowing energy entered your brain core, filled your body with utter bliss and unconditional love, radiated from the core of your heart region, and, afterward, changed your entire life? What would you do after such a neurological "rewiring"? And why not turn to the ancient Indian philosophical tradition that the zapping spiritual teacher taught and that you had just "received" in the little lightning bolt?

I find it fascinating, and hopeful, that this neuroscientist turns to the humanities, and in particular the study of religion, for guidance and help with her life-changing experience and direct knowing.

And she has learned them well. For however close a tradition like Kashmir Shaivism might be to the phenomenological content and practical life results of this particular awakening of a scientific mind, Woollacott does not adopt the Indian philosophical tradition as the final or complete truth. She understands perfectly well that it is another framework, another perspective, one defined now by the first-person perspective of mystical experience, religious revelation, and philosophical discourse instead of the third-person perspective of modern science. We need both perspectives, she insists, within an integral both/and embrace.

Like me, Woollacott is weary of what she calls the "schizophrenia" of Western culture, which rigorously denies the first-person perspective for the sake of the third-person perspective, effectively splitting us in two and resulting in a culturewide dissociative state.[48] None of this requires her to deny a jot of her neuroscientific training and expertise. It simply calls her to reinterpret and recontextualize what it was she was doing for all of those years in the laboratory within a larger cosmic model of consciousness.

Hans Berger searched for a "psychic energy" that could explain his telepathic link with his sister before the rushing cannon carriage. Eben Alexander encountered various light forms of energy in his near-death experience (musical light, the Spinning Melody, the filaments of gold, and so on). When the world "flamed" into life for Barbara Ehrenreich, the cell biologist could not but help invoke the "burning bush" of the biblical story. In each of these, we encounter an altered state of energy that cannot be slotted into the photons and optics of physics.

In Woollacott's little lightning bolt and the currents of love and bliss that it triggered in her body, we can observe how seriously she takes the "energies" that she knew so intimately in the original awakening and in her many subsequent meditation experiences. She knows perfectly well that these are real energies, that they are not simply metaphorical. In a word, they are *empirical*, even if they cannot be made to appear on call in a laboratory setting.

Precisely because she is a scientist, Woollacott knows that this "subtler-than-subatomic reality" cannot be fit into any of our current models from neurology or even physics.[49] These altered states of energy were coded. They contained information. Indeed, they contained an entire philosophy of mind and matter. They were also transmitted from one human being to another, quite intentionally, within a specific ritual context. The energies *knew* what they were doing. They were conscious and intelligent. And they worked with her body and its anatomy in ways that astonished her. Indeed, they revealed to her a kind of second body that was energetic in nature and that was localized in the heart region, as well as in the skull. Because she knows her neurology, she also knows that these energy centers and knowing currents are not employing the nerves and pathways of the physical body. They are doing something else or more.

The possible implications of all of this for fields like human anatomy, medicine, and physics are obvious enough, if, of course, never rigorously pursued. Why would you pursue something that you believe does not exist?

## Michael Shermer

Sometimes even professional skeptics flip, or come close. Michael Shermer is an eloquent and honest man. He is probably America's most famous skeptic, as he is the editor of *Skeptic Magazine* and has a regular column in *Scientific American*. To his great and lasting credit, Shermer did not hesitate to report in his column for *Scientific American* in October 2014 a paranormal experience that he witnessed firsthand.

The story involves a dead radio in a drawer that came to life to play the exact romantic love song at exactly the right moment in order to convince his fiancée and soon-to-be wife (they were getting married within minutes in the same house) that her beloved dead grandfather was present for the ceremony. He writes that this "eerie conjunction" of the dead radio playing just the right song at just the right time "rocked me back on my heels and shook my skepticism to the core." Shermer recognizes that he could come up with rational reasons for why the radio suddenly came to life at that time, but that none of this would or could explain the precise timing of the particular song. Shermer recognizes, in my own words, that the force and power of the paranormal do not lie in any posited mechanism (a battery or wire being suddenly jarred into place, for example). They lie in *the meaning and human emotion* that the event effortlessly expressed. Put a bit differently, such paranormal events have little to do with the methods of conventional science and everything to do with our humanity and, at least in this case, with love.

I confess I am amused by this story and the way it relies on a literal radio, one of the classic metaphors or analogies for the transmission or filter thesis of consciousness. I do not want to dwell on this minor event, and I am certain that it did not turn Shermer into a paranormal enthusiast. It is simply one more signal or sign that the world is not what we think it is, that it does not work like we think it does (as a dead machine), and that at any time *anyone* can be turned around to see this, if only for a brief moment.

• • •

One could write a very big book, even a small library of books, on all of the scientists, engineers, and medical professionals who have either reported robust anomalous phenomena or found them to be of extraordinary scientific significance. If we add to these the tales of entirely secular people, who experience the same but also feel no compulsion to give them supernatural glosses, the library grows even larger.[50] Again, the notion that the mystical and the paranormal are somehow beyond the pale of secularism or somehow irrelevant to the history and practice of science is simply rubbish. Here are a few moments in that history and practice.

Much of nineteenth-century Spiritualism and modern esotericism can be traced back to a scientific prodigy, Emanuel Swedenborg, the Swedish engineer turned clairvoyant and visionary seer who was the assessor for the Royal College of Mines before he was overwhelmed by visions and spent the last quarter of a century writing furiously about them. Kant, recall, (privately) considered him to be the real deal.

Marie Curie was the first woman to win a Nobel Prize and one of the rare individuals to win such a prize twice, in two different fields no less: physics and chemistry. What is not so well known is that Curie also attended séances, including one in Paris in 1907 with the famed psychic Eusapia Palladino, during which the group witnessed luminous points of light forming into a halo around Palladino's head and a bizarre luminosity that could be transmitted to others, including into Marie's hair. Nor were these purely tangential pursuits for the sake of some distracting entertainment or simple curiosity. In the words of Jason Josephson-Storm, Curie "was conjuring ghosts or studying paranormal manifestations as part of her physics research."[51]

She was not alone. Wolfgang Pauli was a pioneering quantum physicist around whose presence poltergeist phenomena erupted regularly and about which the physicist was quite proud. These bizarre mind-over-matter effects were so well known within the physics community that his fellow scientists gave them a name, as if to explain what they could not, in fact, explain. They called them the "Pauli Effect." Pauli engaged C. G. Jung in a quarter-century friendship and correspondence, partly to work through them.

Alan Turing, the great British mathematician, codebreaker, and computer pioneer was convinced that telepathy is real: "These disturbing phenomena seem to deny all our usual scientific ideas. How we should like to discredit them! Unfortunately, the statistical evidence, at least for telepathy, is overwhelming. It is very difficult to rearrange one's ideas so as to fit these new facts in."[52] That is an understatement.

Then there was the British engineer Douglas Harding. While walking in the Himalayas, Harding found himself "without a head"—that is, he discovered that consciousness is not the same thing as the brain or the body at all. "Somehow or other I had vaguely thought of myself as inhabiting this house which is my body, and looking out its two round windows at the world." Not anymore. "I see the man over there [in the mirror] . . . as the opposite in every way of my real Self here. I have never been anything but this ageless, adamantine, measureless, lucid, and altogether immaculate Void." He was being perfectly serious. Douglas Hofstadter and Daniel Dennett would later anthologize Harding's compelling case.[53]

Edward Kelly is a Harvard-trained neuroscientist who became fascinated with parapsychology after his sister became a medium and he encountered a man in J. B. Rhine's Duke laboratory whose psychical capacities spiked off every chart.[54] He spent the rest of his scientific career pursuing these subjects, until he became convinced that, in the last analysis, mind is irreducible and simply beyond the range of any strictly materialist interpretation. He also became the main force behind a fifteen-year symposia series at the Esalen Institute on the brain-mind relationship and the two major books that came out of it: *Irreducible Mind* and *Beyond Physicalism*. Together, these volumes remain the single most substantial and provocative challenge to conventional materialist interpretations of science that we possess.

Then there is the Canadian neurobiologist Mario Beauregard. After hundreds of pages of neuroscience discussing his thesis that the brain is in fact a transmitter and not the producer of consciousness, Beauregard reveals the deepest source of this scientific hypothesis: a classic mystical state that he underwent around 1987 during a serious illness, within which he "merged with the infinitely loving Cosmic Intelligence (or Ultimate Reality) and became united with everything in the cosmos." This state, he explained, transcends the entire subject/object structure of our ordinary experience and is outside of time.[55]

Paul Marshall is a contemporary writer with extensive training in both the natural sciences and the history and philosophy of modern mystical literature. His advanced studies of the physics of motion and Einstein's special relativity theory were partly inspired by a fantastic dream experience that resulted in what he describes as a "vast expansion of consciousness," after which he "looked out at a universe of light. The world was not distinct from me but existed as the contents of my own expansive mind and also as the contents of other minds that I could discern in my vicinity."[56] At the time, it was all hopelessly confusing. He had no conceptual tools with which to understand or integrate this overwhelming experience that "world is mind." He did not yet know that he had been flipped over into a form of idealism. It would be years before he discovered the classic idealist philosophers, especially Leibniz, and could articulate all of this.

Only gradually, then, did it become obvious to him that the two great puzzles of his life—Einstein's special relativity theory and mystical states of consciousness—were perhaps related in some real or actual way. They were just too close. The physics reminded him of his mystical experience. The mystical experience reminded him of the physics:

> The special theory of relativity has led some think-
> ers to speculate that past, present and future events
> coexist in a unity of space and time called "space-
> time." Likewise, mystical experiences sometimes give
> the impression that past, present and future exist
> together in an "Eternal Now." This was certainly true
> of my own experience.[57]

And on and on we could go. These individuals represent, in fact, just a tiny sliver of an immense crowd of professional intellectuals who have not only sensed but have *known*—directly and outside of any sensory or cognitive processes—the fundamental truths of which they speak and write with such conviction. We need not take any of their visions and voices as absolute or final. We would do well, though, to listen to these voices, even and especially when they struggle (and finally fail) to make sense of what happened to them.

Together, their collective presence warns us, loudly and definitively, from any and all lazy assumptions about how the projects of science, technology, medicine, anomalous events, and rare but real forms of consciousness are unrelated. The sooner we understand that, the more likely we—as a community and culture now—will be able to move forward, into new forms of future knowledge and, more fantastically still, new forms of the future human.

# 3

# CONSCIOUSNESS AND COSMOS

*Some physicists would prefer to come back to the idea of an
objective real world whose smallest parts exist objectively
in the same sense as stones or trees exist independently of
whether we observe them. That, however, is impossible.*

—WERNER HEISENBERG

A chorus of cosmic voices.

In his 1980 best-selling *Cosmos*, the American astronomer
and science diplomat Carl Sagan (1934–1996) famously observed
that we are the local embodiment of a Cosmos grown to self-
awareness, that we are "starstuff pondering the stars." Sagan was
referring to the astrophysical fact that the heavy compounds that
make up organic life, including our own bodies and brains, were
fused in the incredible heat and pressure of dying stars before the
latter exploded and seeded space with these same heavy elements.
The line has often come down to us in a more popular and less
sophisticated form: "We are stardust."

The idea is not exactly original with Sagan. The pop singer Joni
Mitchell had beaten the astronomer to the punch ten years earlier

with her 1970 song "Woodstock," which included this same idea in a haunting refrain.

There are other more recent versions of this cosmic sensibility available to us. In a recent YouTube video, the cosmologist Neil deGrasse Tyson, Sagan's modern-day successor, has more recently expressed what he calls the "most astounding fact" that the atoms that make up all of life on Earth, including our own bodies, are traceable to the stars, which, as he puts it, "cooked" the lighter elements into the heavier elements that now make up all life. Yes, we are a part of a shared physical universe. Yes, we live in this universe. But most astonishing of all is the fact that "the universe is in us." This astounding truth, he explains, makes him feel "big" and "connected" as he looks up to the night sky. Is this not what we all want? It is Tyson's nonreligious answer to the religious impulse.[1]

Not every contemporary scientist is so secularizing. Consider the Cambridge paleontologist Simon Conway Morris, celebrated for his pioneering work on the Cambrian explosion of life recorded in such abundance in the fossil record. In his *The Runes of Evolution*, Morris sets out to show, in the words of his subtitle, "how the universe became self-aware." Life and sentience are not some random accidental events for the paleontologist. As the fossil record dramatically demonstrates, they arose many times in many places and show every sign of being woven into the very fabric of things. The thought is difficult to avoid: The universe evolves toward consciousness, toward eyes that see and minds that perceive and, eventually, come to know that they know.

But that is only the beginning. Morris pulled no punches in a public lecture he gave at Princeton University in May 2013.[2] The first punch was in the title: "Nine Evolutionary Myths: The Closing of the Darwinian Mind." It is difficult to listen to the lecture and not conclude that the subtitle is meant to work on two levels: as in "the end of Darwinism" and "the close-mindedness of strict Darwinians." The lecture includes a discussion of convergence as strong evidence of some sort of deeper structure to biology that is not random and ends with some deeply personal reflections on the possibility of the mind-brain as an "antenna" of consciousness (there is the radio metaphor again), the latter suggested by the seeming presence of "a universal music out there" and a Platonic model of the "unreasonable effectiveness of mathematics." The human mind can understand the mathematical structures of the universe because there is something fundamentally transcendent ("out there") about those mathematical structures and, by implication, about the mind that has evolved to pick up and play such "universal music."

Morris ends his lecture as he ended *The Runes of Evolution*, with an attempt to "unhinge" us with stories: "because I am so close to retirement it doesn't matter." Morris told two such stories at Princeton, but there are five in the epilogue, and they are doozies. There we encounter five human beings with impossible abilities or powers, like the mathematical revelations of the Indian mathematician Srinivasa Ramanujan (1887–1920), who attributed his mathematical discoveries not to any simple cognition or problem solving, but to his family deity, the goddess Namagiri of Namakaal, who would write formulas on his tongue and bestow mathematical intuitions in his dreams.[3] Another story, which he told at Princeton as well, is about a time-shift vision of a British air marshal who enters some strange clouds and sees five years *into the future*. A third is about a Catholic priest and friend of G. K. Chesterton, Father John O'Connor, who reported a story about a mental patient in a lunatic asylum who appears to have levitated up to a window in order to escape. As Morris quipped to his Princeton audience, "Again, this will not go down well with the materialists, but, as I said, too bad." The very last lines of *The Runes of Evolution* playfully but seriously suggest that such impossible powers and capacities may well be where the adventure of evolution is heading, that "the journey has only just begun."[4]

An Indian mystic who gets his world-class math from a goddess. A pilot who flies into the future. A madman who floats up to a window in a mental hospital to escape. We might as well be in an X-Men film here—that is, among a secret school of mutants upon whom evolution has bestowed particular abilities and gifts that can only be shamed and feared by our present order of knowledge.

### *From the Classical to the Quantum*

Similar cosmic and superevolutionary sensibilities have, in fact, been expressed countless times before in human history, significantly, almost always in religious or mythological contexts stressing some radical metamorphosis or metaphysical transformation. Some of the earliest origins of human awareness, reflexivity, and symbolic expression, for example, show every sign of having arisen with the origins of religious belief, the special powers of shape-shifting shamans (derisively called "magical thinking" in scientistspeak), and the fundamental conviction, expressed in innumerable mythologies, that the gods, who usually look remarkably human, came from the sky. The message is more than a little jumbled for the modern mind-set but nevertheless not too difficult to discern: We came from the stars.

I do not mean to equate these apparent prehistoric religious sensibilities (whose meanings, pace the popularizers, are not at all clear or obvious) and later historical mythologies of the star gods with the discoveries and mathematical models of the sciences. Disciplines like astrophysics, cosmology, and evolutionary biology have given us in exquisite and truly astonishing detail fundamentally new visions and understandings that no previous culture or era did. Modern science is not ancient religion.

Still, I am curious about these strange resonances between the objective picture of the sciences, "from the outside," and the subjective intuitions and actual experiences of the religious prodigies and their myths, "from the inside." Most of all, I am deeply curious about the philosophical claims of figures like Sagan, Tyson, and Morris, which I find irrefutable, that *we are the cosmos become aware of itself;* that there is some uncanny relationship between the human mind "in here" and the evolving universe "out there"; that, in Tyson's language, the universe is in us as much as we are in the universe.

Although in a figure like Sagan or Tyson such a cosmological claim is expressed in modern scientific and secular terms, these observations bear all sorts of deep historical and conceptual connections to earlier religious thought, and to mystical and esoteric thought in particular. This should not surprise us. If the human mind can mathematically map and so intuitively grasp something of the whole cosmos—and it clearly can, as modern science gives witness again and again—then this is likely so because there is something about the human mind that is itself a reflection (or a reflector) of this same cosmos.[5] The mind is a kind of mirror.

It is not difficult to see why the image of the mirror (which itself produces images) is so attractive. Albert Einstein memorably quipped that "[t]he eternal mystery of the world is its comprehensibility." He even called this comprehensibility a "miracle."[6] What he really meant, I think, is that the most miraculous thing about the universe is that *we* can understand it; that there is something about the human mind that can reflect, uncannily, inexplicably, the hidden mathematical structures of the cosmos, from the galaxies to the garbage in the garage (in my garage anyway). But *why* is this the case? Put a bit more precisely, *how*, exactly, is the human "mind" related to the "matter" of the universe such that it can arrive at such precise, if never perfect, mathematical models of the cosmos? How to explain the inordinately successful nature of mathematics and modern science? How to explain why the universe is understandable at all?

Significantly, Einstein equated this cosmically attuned mind with a most basic and most profound religious sensibility, a "peculiar religious feeling" specific to the scientist, but one that had journeyed very far from its historical roots in figures like Copernicus, Galileo, Kepler, and Newton.[7] Each of these men understood the mathematical laws and geometrical structures of the world to be ultimately located in the mind of God. Mathematics and geometry were literally a kind of revelation for such founding figures. Mathematics was the language of the "book of nature," just as the sacred words of the Bible were the language of the "book of Scripture." Both books revealed the nature and intentions of God.

Such sensibilities would not survive. Thinkers after Newton would soon declare that the math and the mechanics did not need the concept of God to work; that all of that was unnecessary and unconvincing anyway; and that from here on out it was *only* knowledge of quantities that would count as real knowledge. Such attitudes were already in place among intellectuals in the eighteenth century, such as the Scottish philosopher David Hume, who wrote these famous lines at the end of his *An Enquiry Concerning Human Understanding*: "If we take in our hand any volume—of divinity or school metaphysics, for instance—let us ask, *Does it contain any abstract reasoning concerning quantity or number?* No. *Does it contain any experimental reasoning concerning matter of fact and existence?* No. Commit it then to the Flames: For it can contain nothing but sophistry and illusion."[8]

In effect, such thinkers had accepted Descartes's earlier splitting of reality into two separate realms (the mental and the material) but then completely ignored one half of it as unimportant, even nonexistent. Newton's mechanics worked wonders with the "Quantity or Number" thing but said absolutely nothing about the mental world (even though Newton himself was obsessed with theological, occult, and esoteric matters). The eventual result was an entirely objective, mechanical reality completely devoid not only of God but of any mental or conscious dimension whatsoever, a reality entirely independent of any conscious observer or personal dimension. From here on out, such thinkers would focus only on "developing mathematical descriptions of matter as pure mechanisms in the absence of any concerns about its spiritual dimensions or ontological foundations."[9] This was what would eventually become the reigning "positivism" of scientific orthodoxy, which came down to this: The full truth about the material world resides only in mathematics, meaningful concepts can only be stated as quantities, and no other language is real knowledge. In short, David Hume.

Einstein was a complex figure when it came to these questions. On the one hand, he had a real reverence for mystical sensibilities as the deepest drivers of both religion and science: "The most beautiful and most profound emotion we can experience is the sensation of the mystical. It is the source of all true science. He to whom this emotion is a stranger, who can no longer wonder and stand rapt in awe, is as good as dead."[10] On the other hand, he clearly rejected any personal conception of God, which he considered "a sublimation of a feeling similar to that of a child for its father" (this is pure Freud).[11] He also rejected free will on the grounds that a "universal causation" has determined the future every bit as much as the past; he did not believe in the immortality of the individual soul; and he insisted that morality is a "purely human affair," with nothing divine about it.[12] The Jewish genius thus rejected all moral and religious law (including, I presume, the laws of the Torah) but took up "natural law" as absolute, which he clearly saw as an expression of "His thoughts"—that is, as the pattern or design in which we can dimly glimpse something of the "Old One." Einstein spoke of this impersonal God as "Spinoza's God," who reveals himself in the harmony "among all people" and in the sublime workings of the cosmos.[13]

This was no casual or innocent invocation. As a young man, Baruch Spinoza was officially expelled by his own local Jewish community for his "abominable heresies" and "monstrous deeds"—not exactly a warm religious embrace of Einstein's intellectual hero. Such heresies involved things like rejecting the Law of Moses and any providential or personal conception of God. Still, Spinoza was a towering intellectual who not only helped lay the groundwork for the Enlightenment but also advanced some of the earliest historical insights that would become biblical criticism. He also famously (or infamously) captured the relationship between God and the cosmos in a three-word Latin sound bite that could well function as the motto of many flipped scientists today: *Deus sive Natura*, "God or Nature." The other side of Nature is God. The other side of God is Nature. Sounds like the flip to me.[14]

In one of his addresses on science and religion—that is, on this Nature and this God—Einstein warned his listeners not to confuse the technical and mathematical successes of science with the assumed truth of the scientists' present interpretations of the sciences, which, in his own language, are little more than temporary "illusions." He openly proclaimed the need for a new religious or spiritual orientation, one inspired, purified, and guided by modern science:

> It is in this striving after the rational unification of the manifold that [science] encounters its greatest successes, even though it is precisely this attempt which causes it to run the greatest risk of falling a prey to illusions. But whoever has undergone the intense experience of successful advances made in this domain, is moved by profound reverence for the rationality made manifest in existence. By way of the understanding he achieves a far-reaching emancipation from the shackles of personal hopes and desires, and thereby attains that humble attitude of mind toward the grandeur of reason incarnate in existence, and which, in its profoundest depths, is inaccessible to man. This attitude, however, appears to me to be religious, in the highest sense of the word. And so it seems to me that science not only purifies the religious impulse of the dross of its anthropomorphism but also contributes to a religious spiritualization of our understanding of life.[15]

He closed the address by suggesting that "the further the spiritual evolution of mankind advances," the further we will move away from fear and "blind faith" and the more we will strive for "rational knowledge."

Of course, we are a very long way from realizing Einstein's "religious spiritualization" beyond all petty personal fears and all scientistic illusions, much less a "spiritual evolution" toward a truly rational worldview. But Einstein at least saw a relationship between consciousness and cosmos on the horizon of humanity in the polished mirror of science.

None of this presumes that there is a one-to-one correspondence between mathematics and reality; that mathematics somehow exists "out there" in an objective external form as some perfect and complete map of reality (even though this is precisely what more than a few scientists and mathematicians have indeed believed, very much in line, I would add, with the ancient Greek mystical and Platonic sources of mathematics).

And even Einstein resisted what would eventually become the new real—that is, the quantum real—on the basis of his faith in the eventual completeness of mathematical theory to describe the invisible structures of physical reality. His famous debate with Niels Bohr, the architect of the now-standard Copenhagen interpretation (CI) of quantum mechanics, performed this profound discomfort for decades, roughly between 1927, the year of the Fifth Solvay Conference, and 1955, the year of Einstein's own death. As Robert Nadeau and Menas Kafatos describe this debate, at stake was the most basic question about whether *any* physical theory could be complete, could display a one-to-one correspondence with physical reality. At stake were the limits of physics to fully map and so explain the real.

Einstein held the traditional position, inherited from Newtonian mechanics (particularly the massively influential *Principia Mathematica* of 1687) and the confidences of nineteenth-century science, that physics would eventually constitute such a complete theory. In this Newtonian world, reductionism and mathematical modeling authorized the real in an entirely impersonal way. Bohr fundamentally disagreed with this classical Newtonian epistemology or objectivist model of knowledge. He held the position that the logic of complementarity of quantum mechanics (by which an electron can be treated as a wave before measurement or a particle after measurement, but never as both at the same time), Heisenberg's famous uncertainty principle (whereby one can never fully know the position and momentum of an electron at the same time), and the formal role of consciousness and observation in a quantum experiment (in which the quantum wave's "collapse" occurs only at the moment of observation in the total context of the experiment) render this certainty and completeness impossible *in principle*.

Put a bit differently, it was Bohr's position that uncertainty is not a function of an incomplete theory whose loopholes we will eventually close, which was contrary to Einstein's position. Rather, uncertainty and indeterminacy are woven into the very nature of things. To invoke the lovely paradoxical phrase of Abner Shimony, quantum reality is "objectively indefinite."[16]

In 1935, Einstein and his Princeton colleagues Boris Podolsky and Nathan Rosen published a thought experiment (the so-called EPR experiment) to show the apparent absurdity of what Bohr was claiming through a demonstration that his position demanded that two particles, once they have interacted, would display internal correlations regardless of their positions in space and time. Since this seemed to violate the principles of the theory of relativity (whereby no signal or communication can be made to travel faster than light), there was obviously something very wrong about quantum mechanics. It had to be incomplete.

The Bohr-Einstein debate was later taken up by John Bell in 1964 and expressed mathematically in what came to be called "Bell's Theorem," which appeared to show that quantum mechanics, whether complete or not, must indeed violate either "localism" or "realism." Localism, or the principle of locality, is the notion that there can be no causal influence that propagates faster than the speed of light. Realism is the notion that the physical world is constituted by objects that exist independently of observation. These were the base assumptions of Newtonian physics and, really, the modern scientific worldview. One of them now had to go, or so it seemed. Obviously, the stakes could hardly have been higher.

Bell's Theorem showed mathematically what was at stake, but it could not answer the question about whether Bohr or Einstein was correct. This required empirical testing. Bell's Theorem was later repeatedly tested under increasingly sophisticated laboratory conditions at the University of California, Berkeley (John Clauser and colleagues), at the University of Paris at Orsay (Alain Aspect and colleagues), and at the University of Geneva (Nicolus Gisin and colleagues). The eventual results were unequivocal but also bizarre in the extreme. Put simply, they demonstrated conclusively that Bohr was right and Einstein was wrong. Put less personally and more complexly, they provided empirical confirmation of the phenomenon of "entanglement"—the very phenomena Einstein considered absurd and impossible. In lay terms, these empirical tests demonstrated that particles that have once interacted become "entangled" and thereafter correlate with one another's internal states (like "spin") instantly, regardless of the spatial or temporal distances the two particles have since traveled. Entangled particles form an indivisible whole and cannot be treated as if they were separate from one another.

It was now empirically confirmed: Either localism or realism had to go. While it was theoretically possible to let go of realism (and so edge closer to idealism), most physicists preferred to keep realism and let go of localism. They now began referring to the apparent irrelevance of space and time on the quantum level as "nonlocality."

Nonlocality was a very serious blow, or a world-changing discovery, depending on one's worldview. Indeed, the quantum physicist Henry Stapp has seriously suggested that nonlocality may well be the "most profound discovery in all of science."[17] Profound and *really, really weird*. Indeed, nonlocality and entanglement are *so* strange that even physicists have often resorted to paranormal or sci-fi language to describe what appears to be happening: Entanglement is thus playfully likened to "telepathy," a Star Trek–like "teleportation," or "spooky action at a distance" (Einstein's famous phrase). The physicists are joking, of course, but their language is telling, and probably more accurate than they realize or want to admit.

The philosophical implications of all of this were earthshaking, to say the least. Still, such implications have never really been processed by or integrated into our intellectual culture, much less into our public culture and worldview. This is partially due to the reigning positivism of modern science and the academy, which disallows, even demeans, any form of knowledge that is not mathematical or properly "scientific." The simple truth is that numerous quantum physicists, historians of science, and intellectuals *have* tried to translate quantum reality into human terms. As we shall see, more often than not, they have turned to comparative mystical literature to do so, since that is what quantum reality looks like. Such authors have generally not been well received, *not* because their work is not provocative or helpful, but because of the reigning positivism of the academy, which acts as a great "Thou Shalt Not!" written over the door. It's not about knowledge. It's about the politics of knowledge, about *which* kind of knowledge can be known, about *who* gets to know.

Another reason why the new quantum real has not been effectively integrated into public culture is because it is just too shocking to our ordinary sensibilities and ways of thinking. Consider the following gloss on what Bell's Theorem implies by the philosopher of science Robert Nadeau and the physicist Menas Kafatos after their own fifteen-year dialogue:

> All particles in the history of the cosmos have interacted with other particles in the manner revealed by the [entanglement] experiments. . . from the big bang to the present. Even the atoms in our bodies are made up of particles that were once in close proximity to the cosmic fireball, and other particles that interacted at that time in a single quantum state can be found in the most distant star. Also consider . . . that quantum entanglement grows exponentially with the number of particles involved in the original quantum state and that there is no theoretical limit on the number of these entangled particles. If this is the case, the universe on a very basic level could be a vast web of particles, which remain in contact with one another over any distance in "no time" in the absence of the transfer of energy or information. This suggests, however strange or bizarre it might seem, that all of physical reality is a single quantum system that responds together to further interactions.[18]

Such a gloss is very different from the cosmic starstuff quotes with which I began this chapter, quotes that now look almost banal. The starstuff quips work entirely within a Newtonian physics of external objects interacting in three-dimensional space through time, implying a clear logic of reductionism. Our bodies are made of "elements" fused in dying stars. Accordingly, we can be "reduced" to those elements, whose organization lies entirely *outside* themselves via the reactions of organic chemistry and cosmic evolution. We may be stardust, but we're still dust. We are still really just tiny, tiny bits of dead matter.

Not so here in an entangled universe. In the new quantum real, there can be no reductionism in the classical sense, since the organization of the "particles" lies *inside* themselves, as it were, within that strange phenomenon called entanglement. There is no external communication, no chemical reactions, no information signal "between" this and that particle. There is only an instantaneous response from nature outside of space and outside of time. It is as if *everything is already one thing and is simply responding to itself.*

Maybe it is. But how will we ever know if we will not allow ourselves to talk about this possibility, write about it, take it in *any* direction we wish? We are in a difficult place at the moment. Modern science, and particularly quantum mechanics, has rendered every past public conception of the real, every past conventional religious worldview fundamentally inadequate, if not more than a little silly. And yet the positivism of science has prevented us from offering any viable alternative or new story. In Thomas Berry's language, we simply do not have a stable story at this global moment. We are "between stories."[19] Which is a polite way of saying that we are in crisis.

## *Questioning the Rules of the Game*

Much of this crisis is driven by the rules of the game we are playing at the moment and the manner in which they attempt to shut down or deny fundamental human experiences and forms of knowledge that cannot be slotted into the austere rules of scientific positivism and its fetishization of quantity. For example, consider the predictable rebuttals to my earlier observations about the strange resonances between the apparent philosophical implications of quantum mechanics and comparative mystical literature, particularly the latter's insistence that consciousness is fundamental. Such rebuttals come down to (a) some false conflation with or confusion between consciousness and cognition, or between consciousness and the brain, or between consciousness and social ego or personal self (all conflations or confusions that the mystical literature strongly denies and deconstructs); or (b) some version of materialism or physicalism that wants to deny the existence of consciousness as such altogether. In other words, all the rebuttals come down to an unexamined commitment to Newtonian physicalism and a rejection of any form of fundamental subjectivity or first-person direct knowledge.

It is time we define our terms. For the sake of the present discussion on the mind-matter interface, I will frame *materialism* as the conviction that there is only matter, which is fundamentally devoid of mind or intelligence, and that this mindless matter is arranged and behaves according to the mathematical laws of physics (hence the alternate term *physicalism*). In this interpretation, this same mindless matter comes together through exclusively physical processes to produce more and more complex material stuff, which eventually evolves into even more complex organisms that experience themselves as alive and conscious and can reflect back on and come to know the mathematics and mechanics that produced them—Sagan's "starstuff pondering the stars" again. Such consciousness, however, is in reality illusory, for, deep down, the material world is nothing but mindless matter behaving in strictly mathematical ways.

But here is the elephant in the proverbial living room. We haven't the slightest idea how matter makes the leap from insentience to sentience, from dead stuff to us. Nevertheless, we are asked to believe that, somehow, this is the case. When it is pointed out that such a belief is just that, a belief, we are then told that someday we will "prove" that consciousness is an emergent property of matter, and that the materialist interpretation of science is finally the correct one. We are given the hollow shell of a promissory materialism. We are asked to *believe*.

Of course, mental states are clearly connected to the brain in profound and exquisite ways. Moreover, many, even most, mental states and functions may, in fact, be modulated or influenced by brain processes in some way that we do not yet understand. But none of this leads to the conclusion that consciousness itself is only material in the traditional sense, as in "only matter." Consciousness can be *correlated* with neurological processes, with particular parts of the brain and specific neural complexes. Moreover, psychological complexes like the social ego may, in fact, be functions of the body-brain, much as contemporary cognitive science and neuroscience claim. Still, consciousness is *not* ego, and none of this even comes close to identifying a *causal* mechanism for consciousness as such before and beyond every ego.

Do not be fooled by the pretty pictures. Brain scans on some computer screen (the bright colors of which some conscious human being chose and programmed into the software for the computer, which more conscious human beings invented and built) do not explain consciousness. The simple truth is that materialism is only one possible interpretation of the scientific data and the mathematical models that are used to make sense of this data. Materialism is not a fact. Moreover, a materialist interpretation works so well only in as much as it rigorously leaves out everything that it cannot explain, including individual, subjective experiences. Put differently, *materialism only "wins" as long as it gets to declare the rules of the game.*

Such rules might include: "Nothing is real that cannot be established by the scientific method," "Occam's razor is absolute," "You have to use statistics and wash out every anomaly or outlier," "Extraordinary claims require extraordinary evidence," and "All truth must be falsifiable." When those rules are challenged and described as what they in fact are (assumptions, beliefs), typical materialist responses might be "That is pseudoscience," or "That is magical thinking," or "That is anecdotal." At its worst (and it gets pretty bad), such a materialism can function like a dogmatic religious belief, with elite specialists disciplining or shaming those who dare stray from the faithful fold.

## Candy Sprinkles on the Cake of Science

Why is it that we in the modern West seem to know so much about the cosmos but almost nothing about consciousness? I think the answer is simple. The material cosmos is studied as a collection of observed objects "out there." These objects in space can be measured, and their behaviors can often be controlled and predicted with mathematical models. Not so with consciousness. When we study or try to understand consciousness, what is essentially happening is that consciousness is attempting to become conscious of consciousness. The mirror is trying to mirror, become aware of, itself.

Significantly, the mirror is a classical trope in comparative mystical literature in both the West and Asia. Moreover, key terms in the philosophy of mind and the intellectual life in general—like *reflection*, *reflexivity*, and *speculation* (related to the Latin *speculum* for "mirror")—all invoke this same kind of optics, this same base intuition that, somehow, consciousness and light are related.

We are in this way caught, immediately, in a fundamental paradox that will *never* be resolved by studying more objects. I see no way around it: At some point, we will have to "turn around" and look back into the mirror itself. More radically still, we will have to become the mirror mirroring us. We will have to move from third-person observation to first-person awareness. For this, it is useful to listen to those who have been flipped, and to take their collective witness as one possible key to where we ourselves might look now. What we need is a new way of knowing, a new metaphysical imagination that does not confuse what we can observe in the third person with all there is.

Such assumptions end up expressing themselves institutionally and educationally. Timothy Morton's comment that today the humanities are often thought of as little more than "candy sprinkles on the cake of science" playfully describes our present educational moment and why the humanities are not resourced or funded as robustly as they could be. It also jabs at why undergraduates are migrating into the STEM (science, technology, engineering, and mathematics) fields and out of the humanities, and why parents of undergraduates support such a migration and generally see it as entirely unproblematic.

The flip I am imagining here would *radically*, if also, admittedly, gradually, change these academic and social patterns. It is all about ontology—that is, it is all about what we consider to be real and unreal. The status of the humanities is directly tied (if in complex historical and social ways) to the status of the subject or the state of mind in the consensus worldview of the modern Western world, which is now globalized through the spread of technology and science. In this worldview, it logically follows that the humanities should not be valued because they are concerned with surface phenomena, with things that are not really real, that are nonexistent. *The humanities are studying nothing.*

This same ontological logic (never named as such, of course) is what, in turn, defines and determines the entire intellectual hierarchy of universities and, as a result, the deepest metaprograms of our public culture and policies. Those disciplines that study the "most real" things are on the top. Those disciplines that study the "least real" things are on the bottom. Everyone else is in the middle.

So physics is at the very top, because it studies the most real things there are: tiny bits of (nonliving) matter. Chemistry sits just below physics, because it studies things that are more complicated but still quite dead: chemicals and molecules. Biology comes next because, for all of the field's astonishing discoveries, it still has to deal with the tricky question of "life," which seems completely irresolvable on scientific grounds. (At least there are game rules in place to prevent biologists from sliding into the dreaded conclusions of vitalism—that is, the conviction that, actually, life might be irreducible or real.) The social sciences line up a bit below biology because they use scientific or mathematical methods (which allows them to claim a scientific status) but really study perfectly living and conscious subjects—that is, people. Many social scientists, wanting to be "real scientists," try their best to forget this. At the bottom, of course, come the humanities, because they study living things, who speak back no less. They also study completely unreal things, like subjective states of mind, emotion, art, language, and religion.

### A Report on the State of Mind (and Matter)

Reporting on the state of mind in the academy is no innocent gesture. What one is really reporting on is pretty much everything. Many contemporary philosophers of mind realize this. They have initiated a kind of quiet renaissance or invisible revolution in the technical pages and conversations of the philosophy of mind.

I see five related developments in the contemporary philosophy of mind: (1) pansychism, (2) dual-aspect monism, (3) quantum mind, (4) cosmopsychism, and (5) idealism. I've listed them in the order of what I perceive to be their relative difficulty for anyone committed to traditional forms of materialism. These philosophical positions are all *very* familiar to the historian of religions and, in particular, the historian of mystical literature, where they have been most often expressed, of course, in mythical codes and symbolic expressions, not in the technical language of modern philosophy. Consequently, they all seem plausible to me, and I suspect that they all have something important to say.

I move through these five, then, not to vote for one of them, nor to claim some mastery of the philosophical literature, much less of the sciences that the philosophical literature often invokes, but rather to say, "Let go of all those colliding billiard balls and supposedly separate atoms (and human beings). Let go of your deepest assumptions about what a human being is, what or who *you* are."

### *The Big Question and the Is Question*

The single big question that drives most of the modern philosophy of mind is this: "What is the relationship between mind and matter, and how is this relationship mediated or produced by the brain?" On the surface, this is a remarkably easy question to understand. We experience the material world "out there." We experience ourselves as conscious beings "in here." So how are this "inside" and that "outside" related?

But beware of this seeming simplicity. Not only do we not know what mind or consciousness is "in here." The confidences of materialism aside, we really do not know what matter is "out there," either. As both physicists and philosophers have pointed out again and again, all of physics tells us absolutely nothing about what matter *is*. It tells us a great deal, of course, about the causal structures and mathematical behavior of matter, but nothing, absolutely nothing, about what we might call the "isness" of matter—what it *is* deep down in itself.

The key point is actually an almost century-old insight by now (it was probably first articulated in 1927 by Bertrand Russell in *The Analysis of Matter*).[20] But we need reminding. The philosopher of mind Philip Goff points out with particular verve that since Galileo, the physical sciences have proceeded by putting aside the ontological question (the *is* question) and focusing only on the behavior or structure question:

> Think about what physics tells us about an electron. Physics tells us that an electron has negative charge. What does physics have to tell us about negative charge? Rough and ready answer: things with negative charge repel other things with negative charge and attract other things with positive charge. Physics tells us that an electron has a certain amount of mass. What does physics have to tell us about mass? Rough and ready answer: things with mass attract other things with mass and resist acceleration. All the properties physics ascribes to fundamental particles are characterized in terms of behavioral dispositions. Physics tell us nothing about what an electron is beyond what it does.[21]

There are other ways to put this idea. For Goff, mathematical models like those found in physics always "abstract away from the concrete reality of their subject matter." The same, of course, is true of other fields. Economics and political science, for example, may be filled with mathematical models, but these always "abstract away" from the concrete agents and subjects behind all economic and political activity. So, too, with the material world and the nature of matter: "Physics leaves us completely in the dark about the underlying concrete reality of the physical universe."[22]

The sciences, particularly in their infancy, were no doubt wise to pragmatically ignore the *is* questions. The method has been astonishingly successful, and we now know immeasurably more about the behavior and causal structure of the physical universe than we ever have before. But for Goff, *no* such third-person empirical approach will *ever* get us to an adequate understanding of consciousness as a first-person presence. No experiment can get us there, because "the reality of consciousness is a datum in its own right, a datum distinct from the data of third-person observation and experiment."[23]

Perhaps, Goff suggests, all those abstracting observational experiments have taught us enough now to return to metaphysical questions, what I have called the *is* questions, that we could not answer earlier (since we knew so little about the physical universe). Perhaps we can return now, not just with the scientific method but with some other form of inquiry or knowledge, like the philosophy of mind and our own immediate access to our own consciousness.[24] In short, third-person science may not be able to answer the question of first-person consciousness, but it can make us better philosophers, and we may know enough now about the physical universe and the human brain to begin asking the *is* questions once again.

But here is the catch: Common sense and simple sensory data will not get us there. This is how much of the philosophy of mind has been carried on for centuries—by a minute attention to ordinary cognitive and sensory experience. This won't do any longer.

"Metaphysicians have spent too long, as it were, sitting by a micro-scope trying to work out the relative merits of various theories of the microscopic world, without actually *having a look in the micro-scope*."[25] It is time to look into the microscope. Goff thinks this looking into the microscope involves two sources of data: (1) the findings of the sciences with respect to the causal structures of the material world (particularly quantum physics), and (2) "the direct first-person access each of us has to the nature of consciousness," or what he calls simply "the datum of consciousness."[26]

Before we can guess where such a double approach might take us with respect to the question about the relationship of mind and matter and the role of the brain, it would serve us well to know where we have been. In recent Western intellectual history (in the last half millennium or so), three answers have dominated the discussion, and in more or less this historical order: dualism, idealism, and materialism.

The first, often referred to as substance dualism, was Des-cartes's famous answer. There are two independent substances— the mental and the material—and so two different realms of truth. How they relate is not very clear (Descartes thought it had finally something to do with the pineal gland of the human brain). Early science developed by essentially cordoning off and ignoring the mental realm (connected, of course, with "the soul" and all of the religious implications therein) and focusing exclu-sively on the material realm.

The second answer, idealism—namely, the position that mind is fundamental and matter is a function or expression of mind— possessed major voices in eighteenth- and early-nineteenth-century European thought and late-nineteenth-century Anglo-American thought but has since seen hard days. In fact, many of the giants of the Western philosophical canon were idealists of some sort or another: Spinoza, Leibniz, Kant, and Hegel come immediately to mind, but there were many others. Some of our most accomplished scientists were, as well. The physicist who helped initiate the quantum revolution, Max Planck (1858–1947) put it this way: "I regard consciousness as fundamental. I regard matter as derivative from consciousness. We cannot get behind consciousness. Everything that we talk about, everything that we regard as existing, postulates consciousness."[27]

The third answer, materialism, the position that matter is fundamental and mind is a secondary function or surface expression of matter, arose into prominence in the late nineteenth and twentieth centuries, with the impressive discoveries and technological and industrial successes of modern science, which seemed to imply it. Today, the more nuanced position is called "physicalism," by which is meant the thesis "that the complete nature of fundamental reality can in principle be captured in the vocabulary of the natural sciences," particularly physics.[28] The "in principle" part is important, since, of course, this has not yet been accomplished. Physicalists claim it will be.[29] Nonphysicalists claim it will not be.

Many forms of materialism or physicalism today also work with the doctrines that there can be no value-laden causes in nature and "no fundamental mentality"—that is, the position that, deep down, material reality, which is all there is, possesses no purpose and no mental or conscious characteristics. Anything that appears to possess these features, like us, is reducible to, fully explainable by, and wholly grounded in this mindless, purposeless matter, which is all there is.[30]

To simplify, then, we might say that dualism "solves" the mind-matter problem by simply letting both be, whereas idealism and materialism "solve" the problem by eliminating or demoting one of the two members of the binary and absorbing it, sometimes force-fitting it, into the other. But there is little that is actually easy or particularly obvious about any of this, as the present conversation in the philosophy of mind gives ample witness. Indeed, the "hard" problem of consciousness, as David Chalmers famously dubbed it, is often singled out as *the* most important question in all of modern science. Only the nature of "life" in the biological sciences comes close. And many thinkers suspect that these two questions—"What is consciousness?" and "What is life?"—are, in fact, the same question in two different guises.

To the present day, the reigning consensus of the mind-matter relationship within the sciences and much of the academy, including the humanities, remains materialism in some form or another. When the status of mind or consciousness is recognized at all, it is usually recognized as some kind of unnecessary "emergent" property of matter. Philosophers of mind call this position "epiphenomenalism," since mind here is an "appearance" (*phenomenon*) "upon" (*epi-*) a more basic and more fundamental reality—the brain and its neurons, which themselves, of course, are ultimately made of mindless particles behaving according to strict mathematical "laws."

Not everyone is so sure anymore, though, as no such emergence or laws can explain how we get from warm, wet brain tissue to the three-dimensional movie in which we are all embedded and starring right now. The whole metaphor of "construction" or "constructivism" in the humanities implies (but never admits) a more fundamental reality out of which the constructions are assumably constructed. The metaphor, in short, cries out for some kind of ontology. "Constructed" with *what*? But almost no one seems to follow the metaphor (as we will see, the idealists are the dramatic exception here). That would require a metaphysics or discussion of the *is*, after all, which no one is allowed to even utter these days, unless, of course, it is a materialist one. Back to square one.

In my favorite Sidney Harris cartoon, two scientists are standing before a chalkboard that is filled with elaborate mathematical formulas. These formulas extend to both the left and right of the chalkboard. The phrase "Then a miracle occurs" is all that joins them in the middle. If we simply write "matter" over the left side of the chalkboard, "mind" over the right side, and "brain" over the middle, that is pretty much where we are.[31]

"I THINK YOU SHOULD BE MORE
EXPLICIT HERE IN STEP TWO."

### Five Answers

*1. Panpsychism.* One way to join the two sets of formulas on the cartoonish chalkboard and resolve the miracle of how consciousness somehow mysteriously pops into existence from the organization of allegedly dead, purposeless matter is to suggest that matter is not dead at all.[32] It is alive, and it is alive or conscious all the way down. William Seager, one of the leading philosophers of mind presently advancing this model (who came to his convictions, by the way, in a sudden *Aha!* moment while taking a shower), thus speaks of the ubiquity, universality, and fundamental nature of such "presence" or "sentience." In Seager's poetic language, "the world is awake" and, deep down, "humming" with life and feeling.

This is a metaphysical claim, not an observational fact. It is an *is* answer. Once one makes such a move—that all matter is minded to some degree—there is no longer a matter-to-mind problem. Of course, mind emerges from matter, *because matter is itself always and already minded.* This, in a nutshell, is the claim of panpsychism, literally, the position that "everything" (*pan-*) is "minded" or "psyche'ed" (*psychism*). Such an answer, by the way, is already coded in a way that links "mind" and "life," since the Greek *psyche* referenced both (as well as "soul," by the way). The word *panpsychism* itself dates back to the sixteenth century, to an Italian philosopher named Patrizi.

Panpsychism does not solve everything, though. It does not answer the questions of what kind of consciousness, say, an electron possesses (presumably, a fairly simple one). Or how such simpler forms of awareness combine "up" to more elaborate forms, say, like an amoeba, an octopus, or a human being. More seriously, it does not answer *why* they combine up to form more and more elaborate forms of consciousness.[33] This is the "combination problem" of all forms of emergence theory, of which panpsychism is a version. Finally, panpsychism also still possesses a certain "reductive" feel, in that it seems to require us to understand more complex forms of consciousness as reducible to simpler and simpler forms, all the way down to the particles of physics. It feels a bit like atomism, and so materialism, in another now-conscious code.

A number of elite philosophers are usually mentioned at this point, including and especially Alfred North Whitehead, whose process philosophy does indeed look panpsychic. I might also remind you that Barbara Ehrenreich's mystical experience in Lone Pine, with even the teacups in the shop window as alive, could easily be read as a modern panpsychic mystical event.

What panpsychism signals—namely, that everything is alive—is probably the oldest human philosophy of mind on the planet, if we take "philosophy" in a broader sense than it is usually taken in professional circles—that is, as a uniquely Western form of rationalism and reflexivity that dates back to the ancient Greeks. Indeed, panpsychism, under the nineteenth-century anthropological banner of "animism" (literally, "soulism," as in "everything has a soul"), has long been recognized as the fundamental worldview of most indigenous cultures around the planet, from those in Africa to Latin America. It has also been assumed, with plausible if not definitive mythological, artistic, and folkloric evidence, that prehistoric human cultures held similar animistic or panpsychist worldviews.[34]

As David Skrbina notes, none of this makes panpsychism correct, but it definitely works against the common modernist (and materialist) assumption that panpsychism is outrageous or absurd, and that no reasonable human being could imagine such a thing.[35] It turns out that most human beings on the planet have likely imagined (and actually experienced) exactly this "absurd" possibility. It is the modern materialists who are in the fantastic historical minority, not the animists and panpsychists. What is absurd or impossible in one world is perfectly rational and possible in another.

And then there is the oft-heard charge that some forms of panpsychism and cosmopsychism are somehow "New Agey" or "vaguely hippyish."[36] As a philosopher like Goff refreshingly admits, panpsychism and cosmopsychism *do* actually resonate quite well with many New Age worldviews.

*2. Dual-Aspect Monism.* An alternative to old-fashioned physicalism that has received increasing attention in recent years is to consider neither the mental nor the physical as fundamental but, instead, to trace them *both* back to a shared third substratum or superreality. This is essentially what is done in dual-aspect monism. Such a position has a storied pedigree in the history of philosophy and science, including, in different forms, in the works of Baruch Spinoza, Gustav Fechner, Arthur Schopenhauer, William James, Bertrand Russell, Wolfgang Pauli, C. G. Jung, David Bohm, and Bernard d'Espagnat. More recently, dual-aspect monism has been advanced and developed, again in different forms, in the writings of Hans Primas, Harald Atmanspacher, Christopher Fuchs, David Chalmers, and Philip Goff.

Dual-aspect monism comes in two major forms: "compositional" and "decompositional." Compositional forms begin with plurality and work up to unity: Simpler forms of material sentience or awareness come together until more complex forms of life and consciousness emerge. As I understand it, this was Russell's influential position. Here consciousness cannot be reduced to or explained by physics per se (since, again, that is a purely formal third-person description of objects out there in space), but consciousness can be reduced to the "third" insofar as, in Russell's version, both the mental and the material domains arise out of the composition of "psychophysically neutral" elements.

In some panpsychist readings of this form of dual-aspect monism, the elements of the "third" possess both mental and physical properties. Wherever there is matter, then, there is also mentality or protomentality. In other words, dual-aspect monists can be thought of as panpsychists, but of a very special stripe, since they insist that the most basic "stuff" of reality cannot be understood as purely mental or conscious (as in panpsychism), but neither can it be thought of as purely material or physical (as in materialism)—the ground of reality is *both*, or, more technically, it is *neither*. It is "neutral." We find this all difficult to conceive because we are wired to think in dualistic terms (something is either mental or material). We also find it difficult because when we think of matter, we think of it only in its extrinsic or empirical properties—that is, in the third person. This, as we have seen, is not how panpsychists think of material reality. They think of it in terms of first-person mentality, however basic this might be. There is still a combination problem here, though. How, exactly, do these conscious or protoconscious material features add up to rich and reflexive forms of consciousness that we see in a variety of species?

Decompositional forms of dual-aspect monism work in the opposite direction: They begin with a deep unity and "decompose up" into plurality. Such a model has been advanced in the last few decades in a particularly sophisticated and clear way by Harald Atmanspacher. He begins with the quantum physicist Wolfgang Pauli and his long friendship with the depth psychologist C. G. Jung and their mutual working out of the mind-matter relationship—the psychophysical problem, as they called it. This particular form of dual-aspect monism has much to say about the mediating function of the "third," the psychophysically neutral base reality. In effect, it argues that the unity of reality splits into the mental and physical dimensions of our ordinary experience, but that "before us"—that is, prior to any cognition, sensory event, or brain-mediated experience—the real is neither mental nor physical, since there is no split separating and thus defining them. In the technical language of this model, fundamental reality is decomposed up into the mental and the physical through a "symmetry break."

The fundamental nature of reality, then, does not consist of the mental and the physical (roughly, the answer of dualism). Neither does it consist of the physical alone (roughly, the answer of materialism) or the mental alone (roughly, the answer of idealism). This makes the theory both dualist (on an epistemological level "after" and within human experience) and monist (on a deep ontological level "before" human experience"—that is, prior to the symmetry breakdown into the two aspects). Put differently, reality *seems* to be two (or many) to us, but deep down it *is* really one.

Dual-aspect monists, then, possess an elegant answer to how we arrive at our daily experience of twoness, that is—how we arrive at our common sense of being an interior subject perceiving an exterior reality. Our ordinary experience of being "inside" a body looking "outside" is a result of this deeper superreality being split into two experienced dimensions by the brain-body and its particular adaptive needs and spatial and temporal perspective. For a thinker like Atmanspacher, the phenomenal experience of time itself is also created by this symmetry breaking of the deeper reality. Ultimately, there is no inside or outside, no subject or object. Ultimately, there is no stream of time and no structured space. There is only a fundamental oneness that is neither mental nor material. This is what Pauli and Jung called the holistic One World (*unus mundus*).

Here is a simple graphic illustration of dual-aspect monism provided by Atmanspacher and used in numerous conversations of the same:

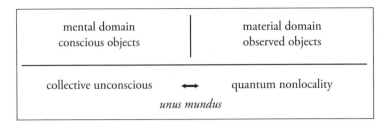

| mental domain | material domain |
|---|---|
| conscious objects | observed objects |
| | |
| collective unconscious  ⟷  quantum nonlocality | |
| *unus mundus* | |

Even if everything in such a world arises due to successive splits, distinctions, or symmetry breakings, the *unus mundus*, or One World, sets this position apart from most forms of panpsychism, as it provides a fundamental cosmic unity from which all local differentiation or plurality emerges. The felt atomism and pluralism of panpsychism is thus answered by a radical monism that is, in the end, inaccessible to human thinking and ordinary human experience (a very important qualification, as we shall see in a moment), since the latter is always an after-the-fact function of the split "up here," well above the One World. We do not just experience this split as the typical subjective-objective structure of our phenomenal experience and grammar. We *are* this split.[37]

As the graph makes clear, there is both a "horizontal" and a "vertical" split in dual-aspect monism, neither of which is final or ultimate (since reality is ultimately One). There is the epistemic split between the mental and the material domains "up here" in the world of human experience: an in-between that we all experience every day. And there is the vertical symmetry break between the ontic ground of all there is "down there" in the One World and the world of human experience of the mental and material "up here." This is an in-between that we normally do not know, *except* in extraordinary "mystical" events that grant particular individuals temporary access to the immanent ground of all being. And even these cannot be classified as "experiences" normally understood (or "particular individuals"!), since there is no subject experiencing an object. There is only the One.

*3. Quantum Mind.* Alexander Wendt is such a dual-aspect monist or panpsychist, with a twist. The twist is that he is an accomplished political scientist and a widely recognized expert on international relations. It is precisely this linking of the physical and the social sciences that makes his work so interesting, and so relevant, here.

In *Quantum Mind and Social Science: Unifying Physical and Social Ontology* (2015), Wendt advances the elegant thesis that most of the dilemmas around agency, subjectivity, freedom, and experience that philosophers and social scientists—and, by extension, I would add, virtually all humanists—struggle with are only problems if we restrict our thinking to what is now an outdated metaphysical world, that of Newtonian classical physics.[38] The mind-matter problem, for example, is simply intractable within any classical material worldview, since it is this classical paradigm and its grammar of individual "subjects" and material "objects" that produce the presumed paradox of the mental and the material in the first place.

For Wendt, there is no such problem once we adopt a quantum mechanical worldview and see human beings as biological organisms that are both classical and quantum in nature and function. Yes, of course, our bodies are classical objects, but our forms of consciousness are not. Mind and social life in particular are quantum mechanical phenomena for Wendt, not classical objects in three-dimensional space. Accordingly, they work like a quantum mechanical wave function (in the Copenhagen interpretation). Such a wave function is a potential reality, not an actual one. Moreover, it is by definition "plural," "paradoxical," and "multiple" (my terms), since the wave function is a set of probabilistic patterns mathematically superimposed upon one another. Until, that is, the same wave function is measured by a conscious agent and "collapses" into a singular material, measurable, empirical event or "particle."

As such, mental and social phenomena, although they remain technically "physical" (since they could presumably be described by quantum mechanical formulas), are also invisible, immaterial, and not amenable to classical ways of thinking. Mind is an expression of the quantum wave function. Matter is an expression of its collapse and observable measurement as a particle. Mind and matter are thus the "same thing" expressed and known by us in two very different ways—that is, within two different types of physics and mathematics.

This quantum turn and its description of human consciousness as a wave function is no fatuous gesture or simple metaphorical invocation for Wendt. Wendt's brother is a physicist. I take it that he has been schooled in and has checked his understanding of quantum physics against his brother's professional knowledge (thereby modeling the kind of collegiality and conversation I am calling for here). Wendt is arguing that human forms of awareness, agency, freedom, and consciousness are *literally* quantum mechanical in nature. We are "walking wave functions." This little phrase is, in fact, the repeated refrain of the book.

Not only that, we are quantum mechanical wave functions that have the freedom to actualize our potentialities (collapse our wave function) in ways that we can choose and partially determine. We are artists and art forms at the same time. "Of course, these decisions are not unconstrained, both internally and externally, but within those restraints the quantum model of man posits an irreducible freedom to create who we are. It is what I take to be an existentialist picture, in which our lives are like works of art."[39]

Wendt addresses any number of subfields here, including quantum statistics, quantum computing, and quantum brain theory, particularly the latter's speculative suggestions about where, exactly, in the brain quantum states might be preserved and not decohere into classical material states (the most oft-discussed suggestion has been that of Roger Penrose and Stuart Hameroff, who proposed microtubules as the most likely locus of quantum coherence). In the end, though, what Wendt is up to is not statistics, computer science, or brain anatomy, but more or less what I read Philip Goff as proposing. He is after a kind of double movement in which the third-person quantum mechanical knowledge of the material world is mirrored by the first-person quantum mechanical nature of human consciousness, a correspondence that, it turns out, scales "all the way up and all the way down respectively." This is not to confuse the mental and the material, but to note their uncanny mirroring and then trace them back to the phenomenon of life itself, which displays both first-person and third-person features ("my claim is that life is a macroscopic instantiation of quantum coherence"[40]), and, ultimately, deep down to a "single underlying reality that is neither mental nor material but from which the distinction itself emerges."[41]

This may sound all terribly abstract, but its implications for the social sciences, for the humanities, and for our political lives are profound. Once we rethink the human "through the quantum," as Wendt does, *everything* changes. I could go on for dozens of pages here explaining how Wendt's quantum model, were it seriously entertained, could instantly transform the humanities and render any number of present debates moot. I am thinking in particular of the present taboos surrounding any discussion of "subjectivity," "experience," and consciousness as such.

Here is how Wendt himself sums up his argument and some of its implications for the mind-matter question in particular:

> [Q]uantum theory admits a neutral monist/pan-psychist interpretation in which "physical" does not equal "material," and instead sees the material world described by classical physics and the mental world of consciousness as joint effects of an underlying reality that is neither. The question then is whether an ontology in which consciousness goes "all the way down" can scale up to the human and specifically sociological level. While there are a priori reasons to doubt it, there is growing experimental evidence that human behavior in fact follows quantum principles. If that evidence continues to mount, it would confirm a key prediction of quantum consciousness theory, according to which our subjectivity is a macroscopic quantum mechanical phenomenon—that we are walking wave functions. That would constitute a basis for solving the mind-body problem, and in so doing unifying physical and social ontology within a natu-ralistic, though no longer materialist, worldview.[42]

Who of us can imagine a deep sociology in which there are and are not true individuals, where every one is a material body and brain but also a nonmaterial wave function that can be "everyone" and "everywhere" at the same time? There are certainly plenty of hints of such a deep sociology in the history of religions.[43]

Classical materialist accounts of human consciousness, and especially religious phenomena, have to deny, erase, and take off the table so much of human experience to retain the illusion of the completeness of the materialist model. Wendt, on the other hand, has elegantly combined both the immaterial and the material, both the invisible and the visible, both the quantum and the classical into a single model of reality, one, moreover, that does not need to drive a wedge between the social and the scientific, between mind and matter, between culture and nature. Not only has he gone a long way toward "unifying physical and social ontology," as his subtitle suggests, but he has also gone a long way toward unifying the sciences and the humanities.

The deepest implication for the humanities is this: Once we take Wendt's quantum turn, the humanities are no longer studying unreal things when they examine states of consciousness from the inside and the outside. Instead, they are studying the interior and exterior expressions of quantum states, which is to say some of the deepest strata of reality itself. They are no longer studying the candy sprinkles on the cake of science. *They are studying the cake.*

Wendt (no doubt wisely) does not touch on this, but it is worth noting that most paranormal phenomena look like quantum phenomena scaled up into the macroworld. I suspect they are precisely that. And I do not mean that in any metaphorical way, either.[44] I mean to describe the real. I mean to make a claim on the cake.

*4. Cosmopsychism.* If Wendt is even close to the truth of things, the implications extend far beyond any academic discipline. After all, *everything* in the universe is defined by quantum processes. Little wonder, then, that Wendt commonly invokes a panpsychist language. If quantum wave functions can be conscious in some way, so can everything else in the universe, including the universe itself.

One of the most common ancient worldviews was what scholars of religion call cosmotheism. This is the idea that the universe (*cosmo-*) is a god (*theism*). The same vision can be found today in any number of figures and movements, including any number of theologians, scientists, and philosophers who have advanced some form of panentheism—that is, the view that the universe or "all" (*pan-*) is the body of God, or is "in God" (*en-theism*).[45] Panentheism is not quite the same as cosmotheism, since the former requires some form of transcendence of or an "outside" to the physical universe, whereas the latter does not. Even if the ancient systems were not always so clear and distinct on this point, the two cosmologies are clearly related.

Today, philosophers of mind and cosmologists are entertaining a similar idea, although they generally do not invoke a "god" or "God." They invoke a "psyche" or some other vague form of consciousness. Goff, for example, has made the serious suggestion that cosmopsychism is a genuine and plausible solution for the problem of consciousness. He questions what he calls the "smallism" that has reigned supreme in the discussion so far—that is, the powerful tendency of philosophers of mind and scientists to reduce everything down to smaller and smaller bits. Instead, he argues for a "priority monism"—that is, the top-down view that the one and only fundamental entity in the universe is the universe itself, and that all conscious subjects are partial aspects of this more fundamental unified subject.[46] The whole is primary; its parts are derivative or secondary.

Goff sees this cosmopsychism as a particular form of panpsychism and distances his own version of it from any notion of deity. Indeed, he observes that there is no reason to think that such a cosmic form of consciousness is not wildly contradictory in content, much less "a supremely intelligent rational agent." Indeed, he sees no reason why it might not be "simply a mess."[47]

5. *Idealism.* Although Goff resists this "temptation," the truth is that the language and structure of cosmopsychism strongly resemble idealism—that is, the position that mind is

fundamental and that matter is an expression or manifestation of some cosmic or universal mind. As mentioned earlier, idealism was once a major position in European philosophy (and it has been central to much Indian philosophy for millennia), but it has suffered a serious demotion in the last two centuries, partly for its clear religious implications (since a cosmic Mind easily morphs into a philosophical God), partly because of the apparent materialist and mechanistic implications of the success of science and technology ("We can make cool stuff, therefore materialism is true"[48]), and partly because of the computer modeling of the brain that came to the fore in the second half of the twentieth century, which tended to reduce everything to a dead machine. Who needs idealism when you have a smartphone?

Still, idealism has not gone away. Very significantly, different versions of it are especially prominent in figures who have been flipped by some extraordinary experience (and I suspect this has *always* been the case, even when we cannot establish it with any autobiographical evidence—for example, in classical figures like Spinoza, Leibniz, and Hegel). The physicist, coinventor of the microprocessor, and technological entrepreneur Federico Faggin is one particularly dramatic case. Since he underwent an overwhelming state of direct mystical knowing, Faggin has been arguing for the primacy of consciousness in the universe and proposing a science of consciousness that is mathematical and rigorous but not reductive or materialist in orientation. Faggin may or may not use the word *idealism*, but his position as articulated online in different videos certainly suggests as much. In one, Faggin presents matter and its mathematics as the "ink" with which consciousness writes its own self-knowing. The physical, mathematically structured world here becomes the mirrored means by which consciousness comes to know and reflect itself.[49]

Faggin is not the only idealist among the computer scientists. Perhaps no contemporary thinker has been more eloquent, more outspoken, and more effective than the computer engineer and artificial-intelligence expert Bernardo Kastrup. Like Faggin, Kastrup emerged from the very heart of modern science and technology. He has worked in major European laboratories, including the European Organization for Nuclear Research (CERN). All credentials aside, he speaks to the digital age on its own terms and against its own obsessions and naïve uses of computer metaphors for understanding consciousness.

Computer modeling is naïve for a number of reasons, not least of which is that it fails to acknowledge the history of earlier technological metaphors for mind. The mind was once a "ghost in a machine," and the universe was once a giant "clock." When will we learn not to take our historically relative metaphors so literally? And when will we understand that *no* point or metaphor in such a history can be an absolute or final one?

Computer models of mind and the artificial-intelligence dreams that they produce derive from what Kastrup calls the "deprived myth of materialism." If this base axiom were true, one could expect sufficiently sophisticated computer chips to become conscious. For Kastrup, the problem is that the materialist claim is not an established fact, but a metaphysical interpretation of the scientific evidence, and one that is exactly upside down: Mind does not emerge as a fragile and temporary product of matter, but matter emerges as a fragile and temporary product of mind. This is why Kastrup is so critical of panpsychism. He sees it as not having broken with materialism, another form of reductionism or emergentism that cannot handle the fundamental nature of cosmic Mind. If panpsychism is a bottom-up view, Kastrup's idealism is a top-down.

In other places, Kastrup suggests that both the thoughts and feelings of the human psyche and what we perceive as matter emerge from some deeper superstructure or symmetry in a universal form of consciousness, which manifests itself at once as human mental activity and the material universe—a kind of Möbius strip of subjective objectivity or objective subjectivity. This may suggest some form of dual-aspect monism. Based on conversations with him, I know that he thinks of dual-aspect monism (particularly the dualism part) as more of a helpful metaphor that may be appropriate enough and close enough to our ordinary experience, but that must eventually be left behind on our way to a deeper and more fundamental idealism.

Recall that quantum mechanics and its later laboratory confirmations appear to show that we must give up localism (the notion that there can be no causal influence that propagates faster than the speed of light) or realism (the notion that the physical world is constituted by objects that exist independently of observation). Most interpreters of quantum mechanics have opted to give up localism, hence all the talk of nonlocality. Invoking recent empirical experiments that appear to refute realism, Kastrup argues against realism. This is especially clear in his blog post for *Scientific American*, "Thinking Outside the Quantum Box," in which he makes the case that "the particles do *not* exist independently of observation"; that "*all* physical quantities—the entire physical world—are relative to the observer, in a way analogous to motion"; and that, in the words of Richard Conn Henry in a 2005 *Nature* essay, "the Universe is entirely mental." For Kastrup, all physical properties or behaviors are "the discernible configurations of a substrate," which is mind.[50]

It is very important to understand that Kastrup did not arrive at his convictions through only rational or scientific means, and that he often expresses his most complex ideas in openly mythical and symbolic ways. He was flipped into his "epiphany" (his word) in ways that he never reveals fully but hints at or gestures toward repeatedly in *Dreamed Up Reality* (2011), a book that also uses computer art, mathematical fractals, and abstract art to express the nature and structure of consciousness (much, I gather, as Federico Faggin might). In his sixth book, *More Than Allegory* (2016), Kastrup describes in some detail how, years later, he came to see how his earlier and ongoing epiphany threw new light on the Christian myth and confirmed or established its idealist implications.[51]

Here, Kastrup explains, we are embodied forms of cosmic mind, split off "alters" in some vast multiple-personality order. These alters have entered God's dream through sexual reproduction and evolutionary biology in order to wake up within the dream, look around the physical universe as the interior of God's brain, and reflect on our own cosmic nature within this same neural galactic network. Kastrup summarizes our cosmic condition this way: "Put in another way, *the universe is a scan of God's brain*; except that you don't need the scanner: you're already inside God's brain so all you have to do is look around. Your perceptions of the sun, rainbows, thunderstorms, etc., are as inaccessible to God as the patterns of firing neurons in your brain—with all their beauty and complexity—are inaccessible to you in any direct way."[52] We are the universe becoming self-aware. We know what God does not know, since we are "inside" God. But—and here is the even more astonishing thing—we have access to what God knows, since we are, in fact, embodied, particularized forms of this same cosmic mind. We exist in, and so can know, both levels of the real.

Here is where the Christian myth comes in. Kastrup's realization of the idealist truth of the same took place in one of the oldest and largest cathedrals of Europe, Cologne Cathedral in Cologne,

Germany, with its Shrine of the Three Kings (perhaps not accidentally, the three famous "magicians" who were "from the East" in the Gospel story). "At once something flipped inside me, like a sudden shift of perspective: I had gotten it."[53]

This radical turnabout involved a meditation on the crucified Christ in which Kastrup suddenly realized that we are all Christs, crucified on the cross of space and time: "we are all hanging from the self-conceptualized cross of space-time, impermanence and confinement. His divine nature is our true nature as timeless mind taking particular, seemingly limited perspectives within its own dream. That Christ is both God *and* the Son of God born into God's creation is a hardly disguised way to express this symbolically."[54]

Kastrup's oeuvre is not simply a series of idealist tracts written by a computer scientist, although that would be remarkable enough in our present intellectual climate. It is an exploration of how cosmic consciousness projects itself into narrative forms, into story, or what we have come to call "myth," and then wakes up out of that same story or myth to know itself not as other but as Self, as Mind, as consciousness as such before and beyond any single ego or organism.

# 4

## SYMBOLS IN BETWEEN

*We may safely predict that it will be the timidity*
*of our hypotheses, and not their extravagance,*
*which will provoke the derision of posterity.*

—H. H. PRICE

Bernardo Kastrup's invocation of the symbol and the myth as expressing something otherwise inexpressibly true, as "more than allegory," goes to the heart of the tension between the humanities and the sciences. The symbol and the myth, after all, are classical categories within the humanities (deeply intertwined with the comparative study of religion). Indeed, a myth in both popular culture and professional science is a falsehood or a lie, not a narrative form of a cosmic truth. Is there any clearer sign of our present dilemma than this double meaning of myth as "cosmic truth" and "lie"?

Now the symbol and the myth, for whatever else they are, are clearly functions of the human imagination. But what, exactly, does that mean? I once gathered a group of about twenty intellectuals, humanists, and scientists alike, around the subject of the imagination at the Esalen Institute, in Big Sur, California.[1] I asked them to do two things in preparation for the five-day symposium. First, I asked each to tell us the story of the strangest thing that ever happened to her or him. Second, I asked each of them to propose a theory of the imagination that would make this strange story understandable, or at least plausible. *All* of them had an exceptionally strange story. *None* of them had a theory of the imagination to explain it.

Now I think I understand why this was the case. One cannot "explain" the imagination, any more than one can explain consciousness, since both are fundamental and cannot be reduced to anything else. This does not mean, of course, that one cannot trace and analyze how the kinesthetic body determines the *content* and *images* of the imagination ("up," "right," "left," movement through space, and so on). But, exactly as we have it in the philosophy of mind and physics, endless discussions of content and structure do not get us any closer to what the imagination *is*.

This sounds suspiciously like the "hard problem" of consciousness. I think that the imagination is an expression, dimension, or function of consciousness itself. This is especially obvious when the imagination becomes "empowered" by "psychic energies" in anomalous states and engages the material world in dreamlike ways. And, indeed, many of the strange stories that we heard that extraordinary week at Esalen sounded *exactly* like the material world behaving in dreamlike "imaginary" ways. We heard about a creepy art object appearing again and again within a complicated romantic situation, despite it being broken and even thrown away. We heard about a honey jar "teleporting" across a kitchen in broad daylight. We heard about a battery in a backpack catching on fire at exactly the right moment (akin to the radio coming on at exactly the right moment in the Michael

Shermer story).[2] And on and on it went. This is the symbolic behavior of a living dream.

If we are ever going to accept, much less understand, the reality and messages of these events, we are going to have to come to terms with the imagination and its power to shape both the mental *and* material.

It is time to admit that a whole series of deep philosophical problems have been haunting our discussion all along. Even when we were honestly struggling with the nature of matter and the philosophy of mind, other basic issues lurked in the background and confused our conversation: questions of *representation* and *meaning*. It is time to draw these issues out of the shadows and shine a bright light on them.

It all comes down to the in-between. We all assume, on some level, that there is a mental world "in here" and a material world "out there," and that there is only one way that knowledge about that material world can get to this mental world—through the senses. In the modern secular world, we also generally assume that any form of knowledge—say in a big dream, a sudden precognitive intuition, or a near-death experience—is not real knowledge, since it is only "in the head" of the visionary. It is a hallucination. It is imagined and thus "imaginary."

This is precisely how the life-altering *Aha!* moments of Berger and his sister, Ayer, Alexander, Ehrenreich, Woollacott, Shermer, and Kastrup become nothing more than "anecdotes," "coincidences," or "interesting experiences" to those who stand outside them. This is exactly how we dismiss, and so miss, the future forms of knowledge. We just assume that there is an unbridgeable abyss between the mental world "in here" and everything else "out there," and that the only things that can get in are those that travel along the tiny bridge of the senses and their technological aids.

But this is not even true in the sciences. After all, *the* privileged mode of knowledge in the sciences is mathematics, an entirely formal or symbolic language that, for reasons that few people dare to think about, actually mirrors and maps the material world in eerily precise ways.[3] And, although the mathematics of the sciences can be checked through experimentation and is related to observation, technically speaking, none of this mathematical knowledge comes through the senses.

Mathematics is "symbolic" knowledge, by which I mean that it is not what it signifies but somehow participates in that which it signifies. No scientist or mathematician believes that the squiggly lines that we call "Arabic numbers" invented by medieval Muslim intellectuals exist as such out there, but all of science proceeds on the philosophical assumption that such numbers *do*, in fact, represent some relatively precise, pragmatically approximate, or reliable relationship *between* a human subject and an objective physical world—that is, between the mental and the material domains. Numbers are symbols that mean something other than what they appear to mean (squiggly lines on a page, screen, or chalkboard) but nevertheless participate in some fundamental way in that which they symbolize (the workings of the physical world) *and* in that in which they appear (the human mind). In this way, they enable the human mind to commune indirectly with the deepest hidden structures of material reality. They join the knower and the known, the mental and the material.

Historically, however, *mathematics is not the only symbolic language that joins the knower and the known.* It is not the only way human beings can intuit and come to know something of the real. And, regardless of our present denials, symbolic processes continue to operate in extremely common human experiences, like dreams, literary, artistic, or intellectual creativity, as well as in more extreme and uncommon ones, like those of our flipped scientists. These are trickier to interpret, of course, and they seem to lack any practical or technological applications, but they remain real forms of knowledge nonetheless, and, at least for those who know them, *they are the most important and meaningful forms of knowledge that a human being can possess.*

I suspect, but cannot establish, that many (not all) of these symbolically mediated forms of knowledge are trickier because they represent encounters with or approaches to the *unus mundus* or deeper ground of all being or, if one is an idealist, with consciousness itself. To take up our graph of dual-aspect monism again, they do not appear to mediate horizontally, from the material to the mental, as in mathematics or ordinary imagining. They sometimes mediate vertically, from the One World "up" to us. Because this deeper ground of being is before and beyond all mental or material form, and is neither mental nor material, it cannot "speak" in human language or even in mathematics. It is of an entirely different ontological order, and so it "speaks" to us in the only way it can: in superstrange images, open paradoxes, and bizarre or absurd narratives.

The very real possibility that "horizontal" mathematical knowledge shares in the same symbolic processes as "vertical" mystical and visionary experience is strongly suggested by all those historical cases in which the two forms of symbolic knowledge appear together and clearly strengthen each other. One could cite numerous ancient examples, like the Pythagoreans and Greek philosophers who embraced mathematics and geometry as mystical or occult in nature. Similarly, the already mentioned twentieth-century prodigy Srinivasa Ramanujan, whose mathematical discoveries are still being proved, clearly saw his mathematical forms of knowledge as related to his religious visions and mystical life.

The numbers zero and one were not just numbers, but symbols of the deepest nature of God, which is to say that he insisted on both a horizontal and vertical dimension to his symbols.

There are also more secular, purely horizontal examples today. Self-described "party-loving jock" and "college dropout" Jason Padgett is such a case. Despite never making it past pre-algebra, after getting struck on the head during a violent mugging, he became a kind of mathematical genius who "sees" geometric and fractal patterns in the natural world, from the flowing of water to the shape of leaves on a tree. He is now interested in number theory and has been described as both a savant and synesthetic prodigy.[4] Here mathematical and visionary forms of knowledge work together in perfect tandem, probably because they ultimately draw on the same deep structures of the human mind. Please note that it was a brain *injury* that released such powers. This sits well with the filter or transmission thesis (since an injury can also be an "opening up") and poorly with the production thesis (since, presumably, an injury would impair, not increase, one's cognitive and mathematical abilities).

It looks very much, then, like the different domains and even levels of reality *can* be bridged by symbolic forms of knowledge that are not strictly sense-based. If this were not the case, most of modern science would be impossible, since there would be no mathematics, and most of the history of religions would have never happened, since there would be no symbolic revelations. Here is where I intervene, as a professional humanist trained to speak and write of the latter religious genres of symbolic expression.

### *From the Sign to the Symbol*

The strange events Michael Shermer experienced with his fiancée on their wedding day were not exclusively dependent on the objective, physical phenomenon of the radio turning on independently. This was an element, of course, but it was not what made the events so eerie. Subjective states and objective events were arising together in order to express *meaning*.[5] That meaning could not be explained by the physical events (the broken radio coming on just then to play that particular song) or by the subjective states (the bride's memories and love for her grandfather). Meaning sparked *in between* the objective events and the subjective states as it seemed to speak of and from an entirely different level of reality—in this case, the world of the beloved dead. It was this correspondence or coincidence and this apparent presence of a dead loved one that rendered the experience so moving and so uncanny for the couple.

But if such an extraordinary event is fundamentally about meaning, if the material world works like a living text or a conscious story in these moments, then it naturally follows that the scientific method is not the best way to understand and engage such moments, since expressions like "living text," "conscious story," and "meaning" bear no scientific sense or operational meaning. Such concepts are not third-person objects that can be observed. They cannot be empirically tested. They cannot be replicated. Nor can they be falsified or proved wrong.

This is *not* a judgment on these events or on the scientific method. It is simply an observation, but one that can prevent a great deal of misunderstanding and a whole bunch of missteps. Those missteps all come down to two fundamental errors: trying to explain something with the scientific method that is not amenable to the scientific method, and then assuming that anything that cannot be so explained must not be real or important, or, worse yet, must be fraudulent or faked.

Obviously, other methods are needed in such situations,

methods that engage texts, stories, and the interpretation of meaning. What are needed are the humanities—that is, precisely those modes of knowledge that work from the assumption that the world, and certainly human nature, often appears as a kind of living text that is communicating with itself through an elaborate web of signals, meanings, and narratives.

This should not be too much of a leap for the scientifically trained mind. The natural world, after all, is awash in semiosis or "meaning making"—that is, in representations and communications through nonverbal, nonlinguistic, nonhuman signals. When my dog Delilah wags her tail, I know that she is happy. When her tail goes down, the hair on her back rises, and she growls as she shows her teeth, I know that she is being aggressive. She is communicating with the world around her, but not in any human language. The same is true when she sniffs every other lamppost, wall, or tree: She is "reading" the "texts" of her world, through her nose and invisible urine traces no less.

DNA is just such a "secret code" or set of communicating signs that appear to determine all of life on this planet. It is not for nothing that biologists chose to call the four repeating elements of the DNA molecule "letters." They understood perfectly well that the molecule is a kind of text, a type of superlanguage that they have spent decades now learning how to decode, read, and even rewrite. But we do not need to be able to understand, or even be aware, of DNA for it to write and read life. Its messages communicate with or without our awareness. Nature writes. Nature reads.

And nature continues to write and read itself within human experience, from the most ordinary sensory cues to the most extraordinary altered states. But if we are ever going to learn how to read what nature is writing in these latter contexts, we will have to come to terms with the structure and dynamics of signification—that is, with meaning itself and how it is transmitted and received within human experience, including and especially in symbolic experience. What we are addressing here is the *real* existence of meaning within a conscious cosmos or "woke world."[6] This is not an easy case to make, since it necessarily works against the opposite and very firm assumptions in both the humanities and the sciences that there is no such thing as real meaning, that all such meaning is nothing more than a human projection or futile wish in a lonely and cold universe utterly devoid of all purpose, intention, and meaning.

The present state of signification or meaning in the humanities is certainly a dire one.[7] And it is of our own doing. The assumption here is that the signifier, of *any* kind, is arbitrary and never truly participates in that which it is signifying; the relationship between the signifier and the signified is entirely external, relative, historically conditioned, and socially constructed. This applies to both ordinary and extraordinary states of consciousness, in effect conflating the ordinary (where the assumptions work quite well) and the extraordinary (where they cause endless confusion).

I have no objection to these convictions as long as they are restricted to the ordinary states of awareness that we inhabit every day as a social ego or named self. This is the "mental domain" of our dual-aspect monistic graph again. To point out the obvious, there are many words or signifiers in human cultures for what in English we code as "tree" or "star," but none of them bears any necessary relationship to the signified—that is, those strange things growing out of the ground or those lovely things twinkling in the night sky on the other side of our graph—namely, in the "material domain."

I have a very big problem when these same assumptions are carried over to the extraordinary side—that is, to specific anomalous states of consciousness. In these cases, all sorts of ontic communions take place, both horizontally and vertically. A human being undergoing a near-death experience or psychedelic trip, for example, may well see and even participate in a three-, four- or even five-dimensional ultrareal world of unconditional love, pure terror, bizarre beings, and neon colors, even trees and stars. But for the standard humanistic reading, none of this bears any relationship to reality itself. It is all "subjective" projection. It is all "imagined." It is all "in the head" of the visionary. Or it is nothing more than a "hallucination" of a dying or "drugged" brain.

This often contradicts the actual experience of the human being who knew such a state firsthand and directly, unlike the third-person skeptic or doubter. What the conventional consensus essentially says to such an astonished or flipped human being is this: "Never mind. Move on. Empiricism is great, until it conflicts with our assumptions." All sorts of "rules" are then invoked to police and protect this consensus.

How did the humanities get here? There are strands and influences that could be identified and followed back into the deep historical past, but most intellectual historians would trace the immediate origins of this revolution back to Nietzsche, who wrote of language as a kind of prison and saw all attempts to claim some stable truth as illusory and dangerous. We are trapped in walls of words, none of which is real. And the biggest unreal word of all is God, which is "dead"—that is, without any real referent and therefore meaningless.

More contemporary postmodern figures, like the French philosophers Jacques Derrida and Michel Foucault, whose dual influence on the present state of the humanities cannot be overestimated, would later follow Nietzsche and write about "language" and "discourse" as an infinite web of crisscrossing references, allusions, and power relations from which the speaker or thinker can never escape. All is "difference," Derrida wrote—that is, every sign or word relies on or refers to some other sign or word for its meaning. There can be nothing that links or joins two things that are different on some deeper level or dimension (as in dual-aspect monism). There is only a "groundless chain of signifiers." In a famous phrase of Derrida's, "There is nothing outside the text."

In such an all-encompassing text from which there can be no escape, any reported real or stable meaning can only be a chimera. Meaning here is *always* a form of "absence"—that is to say, one never really arrives at meaning, since every signifier is always relying on and referring to something else, which itself never arrives. Presence is impossible in the postmodern condition. Roughly, this is what Derrida meant by "deconstruction."

Foucault added a profound moral or political edge to this postmodern form of thinking. For him, these webs of words and concepts are never innocent. They are always structured along invisible threads of power that run, like electricity, through all manner of speaking or thinking and always privilege particular individuals and communities over others. All knowledge is a product and expression of power and the social institutions that enforce it (government, hospital, insane asylum, university, religious institution, prison). There is no such thing as a neutral or innocent form of knowledge. Roughly, this is what Foucault meant by "discourse."

It is easy to be dismissive here, particularly if one has never taken the time to read and grapple with these fierce authors. There is, after all, more truth in their thoughts than most will want to admit.

The gifts of posmodernism are not just truer. They are also more moral, since so much discourse that claims power and authority over us on some eternal ground or sameness (consider "in the name of God" or "America") always ends up creating suffering and injustice for those whose life experiences cannot conform. Such social practices and truth claims are easily deconstructed with these powerful methods. Deconstruction is justice and freedom.

But if the postmodern turn has given us many powerful and important things, things we would be foolish to give up or surrender, it has also taken other things away, rendered them impossible, unthinkable. Foremost among these, I would argue, is the symbol encountered in anomalous events.

Unlike the postmodern sign that only defers, that bears no real relationship to that of which it speaks, the symbol *does* bear a natural, inherent, or ontological relationship to that which it symbolizes (the "other world" or the deeper nature of this world), as well as to that in which it appears and symbolizes (the human being). As with a mathematical formula (only more so), a symbolic expression can join, link, even unite the two poles of the knowing subject and the object known in dramatic ways. A symbol sometimes even appears to work vertically, promising or pointing to a ground of pure presence shared by subject and object alike.

There is a precise logical way to put the difference between the postmodern sign and the symbol. The sign works on the simple principle of what has come down to us as Aristotle's law of the excluded middle, whereby a statement cannot be true and false at the same time, or, if you prefer, a thing cannot be two different things at the same time. This gets translated into the postmodern world as "the signifier can never be the signified."

Not so with the symbolic. This form of knowledge, so

apparent in the big dream, the ecstatic vision, and the mystical experience, works through enigma, paradox, and complementarity. The symbolic expression can be true and false at the same time, or it can reference two different things at once. The symbol communes, even if it never quite identifies or fuses the subject and object, the visionary and the vision. The symbol itself is a bridge over the abyss. It appears *in between*. Whereas, then, the forms of knowledge of our waking consciousness and our present conventional ways of thinking in the humanities are defined by *the logic of either/or*, the anomalous and extreme events that we have been looking at here are generally expressed through *a logic of both/and or neither/nor*.

We do not really speak of symbols any longer in the humanities, because we do not really believe that that cosmos can "speak" to us in this way. But although some are still convinced that the modern world is disenchanted and devoid of living symbols, this is not, in fact, true. And it has never been true.[8] It is not even true among the icons of postmodern thought. Derrida, for example, may be the very epitome of the postmodern moment, but he also became fascinated with telepathy toward the end of his life, considered it to be real, and thought of telepathic phenomena as the ultimate form of deconstruction (since they deconstruct our most basic categories of self, time, and space). And Nietzsche may have written of the prison house of language and shown the hopelessness of grounding any purely rational knowledge, but he also experienced a precognitive dream of his little brother's death when he was young; he wrote of his own literary and philosophical creativity as a real revelation; and he described what he heard from "Zarathustra" (his name for the source of the revelations) in ecstatic and divine terms that are virtually indistinguishable from earlier religious forms. He even claimed the revelation he received about the "eternal return" was of the greatest scientific significance, even though he was also convinced almost no one would understand it. "God" may well be dead, as Zarathustra so famously taught, but the symbol and the revelation clearly are not, since Zarathustra *was* such a symbol and revelation "in between."

## *The Birth of the Symbol*

The contemporary classicist Peter T. Struck has recently given us a profound history of the symbol and its three-millennium run in Western history in *Birth of the Symbol*. Struck locates the birth of the symbol in the Platonic stream of Western thought, whereas what we think of as the artificial metaphor and what I have just articulated as the postmodern sign sit comfortably in the Aristotelian stream. Little wonder, then, that the postmodern sign follows Aristotle's law of the excluded middle and the symbol does not. There was no such "law" when Plato wrote, much less when the pre-Socratics before him were doing things like incubating dreams, invoking their gods, entering caves to alter their states of consciousness, and experiencing the fundamental Oneness of the universe.

Struck turns to the ancient arts, particularly poetry and mythmaking, for the origins of the symbol. He shows us how the ancient Greeks came to mark these genres with a number of words, the most important of which was *symbola* or "symbols." In Struck's elegant expression, such symbols marked "the limits of their texts," that is, the places where the oracle, the poem, the myth, or the philosophical riddle carried their readers and listeners over into a realm of experience and truth that could not be captured by ordinary thinking or by any simple description.

The word *symbol* itself is significant. It refers to two things "put together" (*sym-ballein*). It originally had nothing to do with literary or religious mysteries. Rather, the term invoked the notions of an "agreement," the token that authenticates a "contract," a "passport," or "watchword," or—much closer to the later meanings—an "ominous chance meeting" on the road signaling some divine intentionality or hidden purpose.

To put a modern gloss on the latter examples, a "symbol" was what Carl Jung and Wolfgang Pauli called a "synchronicity," a meaningful, drop-your-jaw coincidence (or, in Pauli's personal experience, a poltergeist disturbance) that signaled some deep psychological meaning or guidance on the road of life by linking or correlating a physical "objective" event with an internal "subjective" state.[9] The Shermer case was just such a synchronicity or "symbol."

A bit later in ancient Greece, the concept of the symbol would come to be used for philosophical riddles (hence the riddles of Pythagoras or the Sphinx were called "symbols" by about 400 B.C.E.), or the means by which the *daemon* or "personal genius" of Socrates guided the philosopher, much as Zarathustra would guide Nietzsche two millennia later.

With respect to the earliest meanings on record, that of the "agreement" or "contract," Greek businessmen would break a pottery shard into two separate pieces, give one to each other, and use the jagged pieces (which, of course, only fit perfectly into each other) as proof of the identity of the business partners in a later transaction. A "symbol" here was literally a whole that had been split off into two and later recombined to form the whole again.

The logic of the symbol is a logic of complementarity in much the way that Bohr conceived the expression (and represented it on his coat of arms with a Chinese version of the broken and rejoined pottery shards of the ancient Greek "symbol"—the joined yin and yang halves of the famous circle of the Dao). The logic of the symbol can also be read as an expression of a dual-aspect monism—that is, an ontology that assumes a deeper oneness from which a twoness emerges and *between which* the communication/communion/contract takes place. A symbol here is not a message from a source to a receiver, nor is it an arbitrary projection of a human being, but a coded meaning or message that appears between a subject and an object, or between the mental and the material domains.

### *The Flip and the Symbol*

The near-death experience of A. J. Ayer is classically symbolic in its paradoxical structure. Ayer demonstrated this both/and logic in his reflections on the way that his youthful Greek education resulted in the Styx-like river of his hospital vision (the mental domain) and in his brave description of the bizarre beings who control space and time (the material domain). He thus acknowledged how some aspects of the experience were constructed by his education but also refused to explain away or deny the ontological punch of the experience. He seemed to think that real information about the cosmos was being communicated to him (the material domain), even if he also recognized that that information was being reshaped by his own brain and education (the mental domain). The vision occurred in between and yet somehow spoke of and from some deeper or hidden dimension of the real.

The visions of Eben Alexander are similarly baroque and fantastic, even as Alexander's own understanding of these visionary experiences displays a clear symbolic structure. He sees the contemporary near-death experience, including his own, as a modern version of the death-and-rebirth initiations of the ancient Greek Mysteries, one of the major historical contexts of what Struck has called the "birth of the symbol." Indeed, Alexander calls modern near-death experiencers the "new initiates."[10]

He is utterly convinced that what he saw shared in the reality of the other world. It is *this* world that is dreamlike. Alexander understands perfectly well that conventional science insists that all meaning is nothing but human projection, but he knows otherwise now. He may have had to be "dragged, kicking and screaming, into this new world," but he knows that "meaning is real."[11] It is embedded in the very foundation of the world. And visionary experience somehow participates in this foundation of meaning. This is the central claim of the symbol, as well and what sets the symbol apart from the postmodern sign.

We have a clear symbolic logic of complementarity at work, even if such a logic is not always apparent in a particular moment. Accordingly, in some places Alexander appears to read his visions as literally true, as if there really are rolling valleys and immense butterflies, trees, fields, even objects and cities, in the afterlife; while at other times he appears to read them in nonliteral ways. He does both, because *both are true in the paradoxical logic of the symbol.*

Alexander is keenly aware of the apparent contradictions here, but, like Ayer, he bravely insists on what he saw, on what he experienced, whether it makes sense to him or his readers or not.[12] Consider, for example, this passage that appears early on in *The Map of Heaven*:

> When I use the word "heaven" in this book, and talk about it being "above" us, I am doing so with the understanding that no one today thinks heaven is simply up there in the sky, or that it is the simple place of clouds and eternal sunshine that the word has come to conjure up. I am speaking in terms of another kind of geography: one that is very real, but also very different from the earthly one we are familiar with, and in comparison to which the entire

observable physical dimension is as a grain of sand
on a beach.[13]

Such a "heaven" is not just a metaphor or an analogy, then,
although it is that, too. It is an actual geography, a place, but not
a place in this observable physical dimension, nor one that proves
the dogmatic certainty of modern-day Christianity, or any other
particular religion, for that matter. He, in fact, expresses conflict
over the title of his first famous book (*Proof of Heaven*). That was
the publisher's decision, not his. He points out that mystically
inclined individuals from the Jewish, Christian, Muslim, Hindu,
Buddhist, and Baha'i traditions have all found confirmation in his
books.[14] In my own terms, his understanding is explicitly comparative and mystical, not local or religious. He explains that this
heaven is precisely "what makes us human." It is "where we come
from and where we're going." It is our "true country," without
which "life makes no sense."[15] So we have a language that does
not mean what it means but that nevertheless points to or emerges
from a place that is "very real" and is the source of meaning itself.

Barbara Ehrenreich's sensibility around the symbolic is also
especially acute and sophisticated. She does not confuse the science fiction, mystical, and even scriptural language that she feels
compelled to use to describe her anomalous experiences with the
literal truth of what happened to her. She is using science-fiction
language because it works so well, not because it is literally true.
Moreover, she suspects some invisible species behind the traditional religious language and myths, some superior forms of life
that we are simply not evolved to detect. Translation and interpretation—burning bushes and mauling angels—is all we have.

Her experience in Lone Pine was a real communion with another
presence that was communicating to her in dramatically palpable,
energetic ways—that is, through the physical environment and

material objects. The mental and the material were both at work here, and there was no way to separate them. Moreover, according to her, what she encountered and how she encountered it was no constructed metaphor, no human projection, no sign of deferral and absence. This was a *real presence* of extraordinary capacity and power that "spoke" or "burned" through things like teacups in a shop window. Put differently, it was a presence that was neither simply mental nor material, but both at the same time.

Perhaps, though, the most instructive example of a flipped scientist proposing a new understanding of symbol and myth is that of Bernardo Kastrup, who has written at least two books about this very subject: *Dreamed Up Reality* and *More than Allegory*. Because Kastrup inhabits the most radical of the five options listed on page 109—that is, an idealist philosophy of mind—it is perhaps not surprising that his model of the imagination and symbolic communication are the most radical, as well.

It is not just that Kastrup is a bold adherent of the filter thesis in its idealist formulation (which entails that the brain is the localization of mental activity in a broader stream of consciousness); or that, "through mechanisms yet unknown to science, our minds have *direct* access to a largely untapped repository of knowledge about reality."[16] It is that reality itself may well be an expression or function of consciousness as imagination. Imagination may *be* reality.

It is important to be careful here. As Kastrup explains, it is not that nothing is real, or that dreams are the same thing as physical reality. It is rather that dreams and material reality exist on a spectrum, and that sometimes dreams can materialize and matter can behave in dreamlike ways. Dreams and material reality may well be made of the same "stuff," which is more fundamental than either.[17]

There is also a kind of sophisticated sociology at work here, since what we think of as physical reality may well be a kind of collective dream in which everyone's assumptions and beliefs are cocreating a public shared material world that behaves in specific ways—that is, according to the "rules" or "laws" of the consensus reality.

Kastrup is not denying the laws of physics. He is suggesting that they might be reducible to the deeper laws of mind or consciousness as such. Here is how Kastrup himself summarizes his idealist hypothesis with respect to the imagination:

> . . . *in consensus reality synchronization emerges across the imaginations of multiple conscious entities, so to form a coherent shared picture. The constraints entailed by such emergent synchronization may be what we call the laws of physics.* Perhaps the apparently fixed mechanisms of nature are merely an epiphenomenon; an emergent property of the sympathetic harmonization of different imaginations, imagination itself being the true primary substance of reality. Perhaps the laws of physics can themselves be reduced to a fundamental metaphysics of psyche.[18]

This is a perfect expression of the flip, since, in our present scientific materialistic world, the psyche and the imagination are seen as epiphenomena of the laws of physics, not the other way around.

Kastrup's "dreamed up reality" is also an answer to the "observation problem" in quantum physics, whereby observation somehow collapses the wave function and brings a single particle into material reality. This is precisely how all of reality works for Kastrup, if in a much more dramatic and fantastic way. It is consciousness as imagination, in which "conception" and "realization"—or, if you prefer, "observation" and "materialization"—are instantaneous and the same thing. Within his own altered states, at least, *"there seems to be no distinction between the process of perceiving and the process of conceiving."*[19, 20]

Like Federico Faggin, Kastrup thinks that the primary symbolic expressions of reality are mathematical and geometric, which emanate or "unfold" from a deeper level of reality still. This is what he calls "the Pattern," which itself consists of a series or collection of "thought forms." In what I take as a kind of penultimate level of reality to which we have some humble access (if only in flipped states), Kastrup finds what he calls "the underlying Pattern unfolding from the Source."[21]

"Unfolding" (which is what *evolution* literally means) implies no ontological break or ultimate difference between the Source and the unfolding Pattern. Indeed, Kastrup himself uses the language of evolution as a kind of emanation: "Somehow, the way new patterns unfolded and evolved was already entirely encoded in, and determined by, the very shapes, angles, and proportions entailed by previous patterns, so that no new primary information was ever added to the thing as it evolved. The entire story was already fully contained in it from the very beginning."[22] This is the purest of symbolic and mythical expressions, a cosmic story unfolding in which there is no difference between the expression or emanation and that which is expressing or emanating it.

All such thoughts are rooted not in any public reason, but deep in the soil of Kastrup's own first-person experience. He is not just speculating. He is also describing. What this actually looked like to him in his own altered states was a geometric world that expresses information through Christmas ball–like globes, "Kandinsky scintilla" (after the esoterically inclined abstract artist), mandalalike patterns, alien sci-fi landscapes, and fractal patterns, all of which he also illustrates in his book with computer-generated (that is, mathematically generated) images. What such visions and encounters felt like was a kind of "falling in love," a familiar human feeling that keyed him into "what I can only describe as a kind of universal 'tone,' or specific vibration. . . . It then amplified what I was feeling in a kind of sympathetic resonance. . . . It was as though I had tapped into the backbone of the universal pipeline of vibrating subjectivity. I could feel this 'tone' as a gentle but nonetheless strong, full, irresistible hum resonating everywhere."[23] He speculates that, "perhaps all things we see and feel, even we ourselves, are like vibratory ripples in an ocean of a single substance."[24]

Such descriptions resonate with others of similar flipped states. The philosopher of mind William Seager speaks and writes of a panpsychist "hum." Marjorie Woollacott invokes the Indian philosophy of Kashmir Shaivism and its central notion of *spanda*, or cosmic "vibration." Eben Alexander insists that he encountered the humming mantra of "om" within his near-death experience. In this shared language of resonance, tone, vibration, and energy we encounter the quintessential logic of the "symbol," in which expression and meaning share an ontological ground with their source (or Source) and are not really separate. So, too, here, the geometric shapes of the Pattern require "no external semantic grounding" and so "break the circular deadlock science is forever confined to." Or again: "Elemental thought patterns do not need to be explained on the basis of their relationships to anything else; they are self-contained embodiments of their own meaning." Kastrup realizes that this all sounds utterly impossible, if not actually mad. He simply invokes what I have called the flip as the necessary prerequisite to understanding: "How this can be so is something that cannot be satisfactorily explained, but only experienced firsthand."[25]

What was most meaningful to him about it all was the discovery that he is not actually his social ego or "Bernardo Kastrup"; that there is a "transcendent I" behind whom he thought he was but, in fact, is not. It is this "transcendent I," this first-person subjective state of pure presence, that sets Kastrup's thought apart from the knowledge models of third-person science. He knew such a form of consciousness "beyond time and space, which I would normally think of as an abstract 'other person' foreign to me" as "an unambiguous *remembrance* of who I *really* am and have been all along."[26] In one place, he even speculates that this individualized I beyond his present ego is who he was before he was born.[27] It would be difficult to imagine a more Platonic observation.

We have come a very long way: from the imagination as an entirely subjective spinner of fantasy and imaginary schlock to a proposed "physics of the imagination," in which the imagination is the very architect of the universe, with the mathematical and geometric structures of deep reality being its primary symbolic language and direct unfolding or evolution.[28] We have arrived at "the thought language of the Unified Mind whose imagination shapes the unity of all existence."[29]

### *Privileging the Extreme*

Philip Goff writes: "while the success of physical science gives us reason to think that human beings are good at mapping the world's causal structure, it gives us no reason to think we will be especially good at discovering its deep nature."[30] This may well be true with respect to physical science, but there is overwhelming historical evidence that some individuals are very good at discovering what they insist is the deep nature of the world, even if they cannot always speak of that deep nature in mathematical or scientific terms.

That historical evidence is comparative mystical literature, by which I mean texts from around the world and throughout human history that express some exotic altered state of consciousness and energy that claims direct and immediate access, beyond the senses and outside any cognition, to the "secret" (*mystikos*) source or ground of all knowledge and existence: consciousness itself. It is here that I think the philosophy of mind should go if it is really serious about cracking the code of consciousness.[31]

We certainly have a long way to go. Even Einstein was no Einstein when it came to understanding mystical literature. When he wrote of subjects like the ancient pre-Socratics, whose thought was often dominated by the initiatory secrets of the Greek Mysteries, mystical states of oneness, and the magical coincidences of the symbol, he grossly misread them as claiming that "it is possible to find everything which can be known by mere reflection." That is precisely what they were *not* saying. In essence, Einstein confused cognition with consciousness, cognitive thought with direct mystical knowing beyond cognitive thought.[32]

Federico Faggin, on the other hand, not only understands mystical literature; he *knows* his own mystical states, like the pre-Socratics. His case clearly signals what this conversation about the nature of mind and matter would look like if it stopped restricting itself to ordinary mental experience—that is, to the social ego, to things like the qualia of redness or mintiness, ordinary sensory data, and cognitive processes that any computer can mimic.

What would the philosophy of mind look like if it took flipped individuals like Federico Faggin seriously—that is, if it took mystical forms of all-knowing as seriously as it took abstract philosophical reasoning and the color red? As it is, the philosophy of mind is almost entirely restricted to ordinary rationalism and easily accessible conscious states. I simply have no faith that we will ever get to an answer about the fundamental nature of consciousness by "just thinking," or through ordinary states of awareness, much less by abstract reflections on mintiness or redness.

I take science itself as one of our best guides here. In order to advance our knowledge of particle physics, we had to build billion-dollar technologies to expose matter to some of the most extreme and violent conditions in the universe. Only then did matter display some of its deepest secrets in the form of quantum fields, which, in turn, gave us a window into the high-energy states of the early cosmos.

While ethically we cannot expose people to extreme and violent conditions, life does this for us. If we look at what happens to a human being in extreme conditions (like a near-death experience or a traumatic paranormal event), we will likely get much closer to the truth of consciousness than we ever will by talking endlessly about qualia and cognitive modules.[33]

It is no accident that the pioneers of quantum mechanics themselves went straight to mystical literature for the best "up here" analogies to quantum mechanical theory. They did not claim scientific status for these speculative comparisons, but they suggested deep parallels. It is difficult to read them without suspecting that they thought that mystical states were quantum mechanical in structure and nature.

As already noted, it was Niels Bohr who put the Chinese Dao symbol on his coat of arms to capture his famous logic of complementarity. Werner Heisenberg was nicknamed "the Buddha" for his profound interests in Indian philosophy. He was also the first to suggest that the so-called matter of quantum mechanics looks very Platonic in nature (that is, it looks more like a thought than a thing).[34] Erwin Schrödinger had a similar interest in Hindu and Buddhist literature and thought that consciousness was fundamental—that is, not reducible to anything else. Wolfgang Pauli was convinced that some fusion of quantum physics and mystical experience constituted the future of thought. Such convictions were forced on him by his own extensive experience with psychokinetic effects around his own body and person.

How do we make sense of such physics-to-mystics intuitions and experiences? What all such quantum mystical literature has in common with the five models of the mind-matter relationship discussed previously is the understanding that the mathematical behavior and physical structure of reality—that is, the "objective" or "outside" structures mapped by science—must be realized by some inner stuff or substance, even if the math and physics can never tell us what that "something" on the "inside" is. The further bold thought here is that *the realizer of all that structure and mathematics might be consciousness itself.* In other words, there might well be a "subjective" or "inside" to reality that the scientific method can never get to in principle (since it only deals with the "outside" of things) but that mystical experience can and does all the time.

If this bold thought were true, one would expect that the inside of things would reflect or be complementary to the outside of things, that the physics and the mystics would correlate in some parallel fashion, without ever, of course, being confused as the "same thing." This is exactly what the quantum mystical literature has argued now for almost a century.

As the early quantum physicists intuited, the structural features of mystical forms of "mind" and the weird behavior of "matter" in quantum mechanics *do*, in fact, look remarkably alike. The intuition is likely correct, although we must be careful not to push the parallelism too far or take it too literally. Here are some shared, parallel, or coinciding features.

- Like a quantum state, many mystical states of consciousness are *in principle* unrepresentable and indeterminable. Accordingly, they deconstruct any and all traditional images of "God" or the "human." In the technospeak of the study of religion, they are apophatic, a Greek-based term that means "saying away."

- Such states typically engage both mental or spiritual and material or physical domains and effects. They commonly appear as an immanent irruption into human awareness of a pure presence or undifferentiated experiential ground without distinctions and without any subject-object splitting, much as we have it modeled in dual-aspect monism.

- Such states are often profoundly "hermeneutical" or "interpretive" in radical ways—that is, in mystical experience as in quantum mechanics, it is the act of observation or the experience itself that determines the behavior or appearance of the real. This is why the expressions of mystical experience are so different across cultures and times.

- Like quantum processes, it is equally true that mystical states are fundamentally holistic—that is, they emphasize sameness over difference, unity over plurality. Indeed, if there is any common message of mystical literature, it its that everything is one thing, that "all is One."

- Such states are routinely expressed through symbolisms of complementarity or open paradox, things like snakes biting their own tails, sexual union, and the yin and yang of Chinese Daoism. Hence Bohr's attraction to the same.

- Such states are commonly expressed and experienced through the same phenomenon that transformed classical Newtonian physics into quantum physics—the cosmic nature of light. It is not for nothing that mystical states are routinely described as "illumination," "enlightenment," "conscious light," "uncreated energies," and so on. I really cannot stress this enough, although I know it will generally not be well received: I strongly suspect that mystical experiences of light and energy are experiences of light and energy "from the inside," whereas the physics of light and energy is mathematically mapping light and energy "from the outside."

- Such states often express themselves through dramatic "nonlocal" anomalous phenomena, like experiences of eternity (or no time) and the immediate knowledge of objects or events at a spatial or temporal distance.

- Such states often propose two levels of truth, a conventional or exoteric level and an ultimate or esoteric level, much as physicists today make a "two-domain distinction" to discuss the differences between classical and quantum features of reality.

- In a similar vein, such states often result in doctrines of what I have called the "Human as Two"—that is, they posit two forms of human consciousness: one that is entirely invisible and in some sense "transcendent" to any local realism (not unlike consciousness as Wendt's "quantum wave function"); and one that is embodied, historical, material, social, and so on (the human as a classical Newtonian object or embodied and "particularized" material ego).

One could go on here for a very long time and provide endless examples. Allow me just one instructive example: the "block universe" hypothesis of modern cosmology, which repeats in almost perfect detail numerous features of mystical experience, from the simultaneous existence and nonexistence of space and time *sub specie aeterni* ("under the perspective of eternity") to various altered states of consciousness in which the material world is seen to move and not move "at the same time."

Consider one particularly illustrative scene involving the most basic of physiological events: a legendary Chinese Buddhist monk and poet named Han Shan (literally "Cold Mountain") taking a pee in the backyard:

> As one comes suddenly out of darkness, I perceived the full meaning of the doctrine of immutability and said: "Now I can believe that fundamentally all things neither come nor go." I got up from my mediation bed, prostrated myself before the Buddha

shrine and did not have the perception of anything in motion. I lifted the blind and stood in front of the stone steps. Suddenly the wind blew through the trees in the courtyard, and the air was filled with flying leaves which, however, looked motionless, . . . When I went to the back yard to make water, the urine seemed not to be running. I said: "That is why the river pours but does not flow." Thereafter all my doubts about birth and death vanished.[35]

For the contemporary cosmological parallels to such an ancient Chinese scene, all one has to do is go watch the astrophysicist Brian Greene explain the block universe on a YouTube video.[36] He even uses the same image of the "river of time" as a "frozen river" that flows and does not flow at the same time.

It is difficult to read the ancient Buddhist teacher and watch the modern cosmologist and not suspect that they are talking about the same thing in two entirely different cultural codes. It very much looks like we can know reality *as it really is.* Could it be that we are all quantum amplifiers, even if some of us amplify quantum processes more dramatically and clearly than others?[37]

It should not surprise the reader to learn that I am in fundamental disagreement with the oft-heard claim that "quantum mechanical events cannot be directly perceived by the human sensorium."[38] Such a statement ignores a growing body of evidence to the contrary within the new field of quantum biology, including the simple fact that the human eye can detect a single photon, which is quantum mechanical in behavior. It also ignores Wendt's fascinating suggestion that consciousness itself may *be* a quantum mechanical event, which would make all forms of human consciousness quantum mechanical in nature. And then there is the real possibility that there are other "mystical" ways of perceiving and knowing the universe outside and beyond the five senses, as modern intellectuals from Francis Bacon on have repeatedly stated. These latter ways of knowing may not "make sense," but this does not mean they do not exist, only that they cannot be slotted into our ordinary or usual ways of knowing the world, and that they violate our positivistic dogmas. So?

Mystical literature is the place to go, too, because we find here the most extreme and the most telling examples of reflexivity, of consciousness knowing itself as and through itself. This is where the mirror turns back on itself, stops reflecting objects and egos, and comes to know itself not in the mirror but *as the mirror*. It is in this literature that we can see, if still from afar, consciousness consciousing consciousness.

Among countless examples of this in the history of mystical literature, few are more eloquent than the great German medieval mystical theologian and professor Meister Eckhart (1260–1327), who famously captured the nature of God (and consciousness) this way: "the eye with which I see God is the same eye with which God sees me." He also preached of how one has to move beyond what people think of "God" (some big person over there, like that chair) to the "Godhead above God," who is, in fact, a "Nothing," in which there is no space or time, nor any "Conrad" or "Henry"—that is, no personal ego. In short, Eckhart was preaching (literally) a form of consciousness beyond space and time that could see through the social self and embodied ego (the Conrad and Henry) but was no such thing.

Eckhart was a highly trained Christian theologian and philosopher who belonged to one of the most elite religious and intellectual organizations of his time, the Dominican Order. But one does not need Christianity to express the same reflexive structure. Consider Advaita Vedanta, the justly famous "nondual" (*a-dvaita*) school of Hindu philosophy, which has involved a most basic distinction between the immortal Self (*atman*) and the social persona or mortal ego (*ahamkara*) and a subsequent contemplative project "[to know] the Self in the Self by the Self" (*atmani atmanam atmana*).

Zen Buddhist practice depends on contemplating these very paradoxes, paradoxes irresolvable on the plane of the ego and its dualistic, or binary, thinking. Hence its most famous koan or riddle: "What is the sound of one hand clapping?" One could just as easily have asked, "How does the eye eye the eye?" Christian, Hindu, and Buddhist mysticism are not so far apart, not at least here in the paradoxical structure of consciousness itself before and beyond any mental subject knowing a material object.[39]

What separates mystical forms of thought like those found in Han Shan, Meister Eckhart, Shankara, and Zen Buddhism from the contemporary philosophy of mind is that the former mystical expressions insist on the possibility of directly *knowing* consciousness and cosmos beyond or before the splitting of human experience into a subject and an object, whereas the contemporary philosophies of mind, even if they posit such a fundamental reality, do not seem to be aware that such a direct and immediate knowledge is possible at all. Indeed, the contemporary philosophy of mind, and indeed the scientific method itself, work only if the subject is separated from the object. Epistemologically speaking, philosophical reasoning and scientific experiment depend on a kind of dualism in which the subject can never really know the object directly, can never really *get* to the isness of reality. All true knowledge here is indirect or mathematical knowledge. We can never commune with or become that which we know.

The same is not true in mystical literature. Indeed, the direct and immediate knowing of this isness or suchness is the whole point of much mystical literature. I suggest we go *there* if we are really serious about plumbing the depths of consciousness and cosmos, of mind and matter.

# 5

# THE FUTURE (POLITICS) OF KNOWLEDGE

*Once, the cosmos was etched into concentric spheres with God*
*in the middle, a macrocosmic representation of feudalism.*
*Now, geneticists like Dawkins argue that what we see*
*as animal life is really just a capitalist free market*
*in genetic code. Whenever you hear a rapturous*
*defense of the natural world, you should be on your*
*guard: this is class power talking, and it's trying to kill you.*

—SAM KRISS, "Village Atheists, Village Idiots," *The Baffler*

It matters how we think of consciousness and the cosmos. Every model of the natural world carries implications not only for how we understand ourselves but also for how we organize ourselves socially and morally. What are the ethical and political implications of the flip? Do the possible impossibilities of anomalous experience, the life-changing personal revelations of flipped scientists and intellectuals, the corresponding quantum natures of consciousness and cosmos, and the real existence of meaning in the physical world really have any bearing on the day-to-day world in which we all live, love, die, and hate? After one has realized that consciousness is fundamental to the cosmos and not some random evolutionary accident or surface cognitive illusion, that *everything* is alive, that *everything* is connected and in effect "One," then what? Would the billiard-ball selves of the Newtonian world and the political systems and values built around them over the last few centuries make sense any longer?

They would certainly make *some* sense, since we would not cease being persons, more or less identified with singular relatively self-contained bodies and a remarkably consistent form of consciousness that does seem to be individual—that is, undivided. But if we really have been flipped, would we live differently?

Even those who resisted the implications of quantum physics, like Albert Einstein, saw clearly that the new real implied by the new physics also implied a new ethics, one in which everyone was a part of the same whole and not really so different. Here were the ethical and political implications of Einstein's cosmic spirituality:

> A human being is a part of the whole, called by us the "Universe," a part limited in time and space. He experiences himself, his thoughts and feelings as something separate from the rest—a kind of optical illusion of his consciousness. This delusion is a kind of prison for us, restricting us to our personal desires and to affection for a few persons near us. Our task must be to free ourselves from the prison by widening our circle of compassion to embrace all living creatures and the whole of nature in its beauty. Nobody is able to achieve this completely, but the striving for

such achievement is in itself a part of the liberation
and a foundation for inner security.[1]

It is difficult to overestimate the radicalism of such a position,
particularly in our present moment. Could Alex Wendt's new
quantum being really "own" anything? I can purchase objects
with a socially constructed abstraction called "money," but even
when the thing is home and sitting on my desk or in my living
room, or wherever it is sitting, the laws of physics keep me from
actually ever touching it completely (for every touch is not really
a touch, but a repelling of forces on an atomic level). In truth, I
own nothing.

"You can't take it with you." And you don't even really *have*
it here. All you can do *even in a Newtonian world* is get close to
things for a while. The absurdity arises because a "soft" quantum
being (a wave of consciousness, as it were) is attempting to behave
like a "hard" Newtonian object—get close to something, "own"
it, as we say. But what if this wave just stopped trying to do that?
What if consciousness could free itself from the reign of such
objects altogether?

My point here is not to suggest a plan of action for our present world, but to point out that this *can* be done. Consciousness can cease to identify with any and all objects. It is possible,
even if, historically speaking, this has been accomplished only
in elite religious communities and by particularly gifted or rare
individuals.

The history of religions is not of a single voice on the moral,
political, much less the economic results of such attempts and
accomplishments. Historically speaking, the flip does not automatically translate into simple ecological lifestyles and compas-
sionate behavior, much less to a collapse of all difference in a
global community or cosmic spirituality.

Individuals who have been flipped can be sexually abusive, physically violent, racist, discriminatory, and just plain mean. Saint Bernard of Clairvaux (1090–1153) was one of the architects of medieval Christian mysticism who spoke and penned eloquent sermons on the nature of mystical love. He also preached the Crusades. Japanese Zen monks have sought satori, or enlightenment, for centuries. In the twentieth century, many of them were ardent nationalists who encouraged their fellow Japanese citizens to go to war with the Allies. Many forms of Hinduism, including Hindu monasticism, are deeply informed by profound nondual philosophies of mind, yet Hinduism in India also depends on the caste system for its day-to-day social and political expression. The recent history of Hindu and Buddhist spiritual teachers in the West, all claiming some kind of fliplike enlightenment, is absolutely rife with the traumas of sexual abuse and outright moral deception and blatant lies.[2] The conclusion is unavoidable: The flip does *not* equal moral and political enlightenment in our present liberal senses.

• • •

I do not wish to make a political or economic argument about the quantum futility of owning things (although I happen to believe that). I do not wish to argue that the flip is some kind of sure-bet way to become a moral and politically enlightened person (I certainly do not believe that). Actually, I do not want to argue that a personal experience of the flip is necessary at all.

I want to make a different, more practical, and more humble case. I want to argue that, even in their present distant secularized forms, the humanities carry something of their original spiritual impulse (they were born in the classical and magical interests of the Renaissance, in the theological projects of the European monasteries, and, further back still, in the mystical impulses of the Platonic Academy). Accordingly, they carry a certain flipped structure that can well be called "prophetic" in the ways that it calls out social injustice of every kind and does not depend on any single person or charismatic figure. Consequently, an engagement with the humanities in systematically careful and rational ways— that is, through public and private education—remains the best way to translate the flip into sustainable social, political, moral, and economic forms.

## *The Prophetic Function of the Humanities*

One primary function of the humanities (though not the only function) involves exposing, analyzing, and criticizing the unjust structures of human society. The humanities as a whole serve a prophetic function in society, and this is precisely why they are often resisted so, even presently hated under the banner of the "liberal professor." One can come up with all sorts of good and bad reasons why the humanities should be challenged, but behind most of them (good and bad alike) lies a disguised attempt to preserve and defend unjust social structures and practices, often of a racial, gendered, or class-based kind.

Some very basic skill sets burn at the heart of the prophetic witness I am imagining here: (1) reflexivity, or the ability to move outside one's own world and observe it, critically and compassionately, from the outside; (2) fair and just comparison of other peoples and other communities; (3) a fundamentally different spiritual orientation or "religion of no religion" that locates the locus of fuller religious truth in the unrealized future and not in any imagined golden past; (4) a cosmic humanism that understands the human as an expression of the entire universe; and, finally, (5) a deep, dark ecology understood as self-care.

As with my summaries of the new (or ancient) philosophies of mind, I list them here in order of what I perceive to be their existential difficulty. In the humanities, the truths discerned almost always offend or violate the status quo and the comfortable, some more so than others. That is *precisely* why these truths are so important, and why colleges and universities are so important—because these are the only institutions that nurture and support professional humanists and intellectuals in large numbers. These institutions fill a vital function of any healthy modern society: the function of robust and unflinching self, social, and political critique.

## 1. "Don't Believe in Yourself"

The most basic method of the humanities is reflexivity, which is the ability of thought and awareness to turn back on themselves in order to think about thought and become aware of awareness. Once we become sufficiently reflexive, it becomes painfully apparent that our deepest convictions, beliefs, thoughts, even emotional reactions and sensory impressions are not what we thought they were. We are not cameras taking neutral photographs of the world as it really is. Rather, our impressions, responses, and perceptions, our deepest beliefs about the world and ourselves, are constantly being shaped, redirected, and distorted by all sorts of influences, of which we are generally completely unaware—neurological, cognitive, and linguistic structures, but also deep historical trajectories, gendered, racial, and class assumptions, learned cultural values, economic privileges, structures of poverty, psychosexual patterns, base religious metaphors, and so on. The reflexive turn is a very humbling realization, but it is also a kind of secular spiritual practice, since such reflexivity allows one to "step back" or "step away" from oneself.

Put succinctly, *reflexivity is an intellectual form or expression of the flip* in which thought turns back on the thinker and examines that thinking subject critically. Because of this unique "looking back," reflexivity is also the ability *not* to think one's thought, *not* to believe one's beliefs, to realize that one is not the thought or the belief or the social identity or anything else that has been constructed. Reflexivity is the call not to believe in oneself, at least not the ego. Reflexivity is a kind of rehearsed or ritualized out-of-body experience.

This secret of reflexivity, I suspect, is another reason that the humanities are feared and misunderstood by those who cannot or will not acknowledge a deeper humanity or form of awareness behind the masks of religion, culture, and ethnic group—that is, *behind themselves.*

The humanities dissolve such illusions like the morning sun evaporated the fog of the low-lying creeks and riverbeds of my Nebraska boyhood. This (and the counterculture) is why higher education has tended toward a liberalism or progressivism over the last century or so. This is not a function of some kind of conspiracy or desire to suppress conservative voices. Disciplined and nuanced conservative voices are most welcome and, indeed, often nourished by that very education, and conservatism constitutes a precious critique of the kind of intellectual fads to which academics are particularly prone. Often, the liberal intellectual can *only* dissolve, can only deconstruct. This is where the conservative criticisms of contemporary liberalism are quite accurate and insightful, and why liberal intellectuals need to listen to them very carefully now.

## 2. *"Comparison Is Justice"*

When neo-Nazis, KKK racists, white nationalists, and alt-right activists marched into Charlottesville, Virginia, on August 11–12, 2017 under the banner of "Unite the Right" and shouting "The Jews will not replace us," they marched with tiki torches that were designed to remind black people of the white terrorism that was central to southern culture, American slavery, and Jim Crow laws. They also marched with the moral support of the president of the United States of America and the silent blessing of a disturbingly large portion of the American population. The broad silent support of the evangelical leadership and voting bloc was particularly telling, and morally repellent.

There is always a "logic" behind such hatred, even when it is a false and twisted one. When Trump compared the neo-Nazis and white nationalists with the counterprotesters—his infamous "on all sides" line—he engaged in a classic case of bad comparison. He equated two things that are exact opposites (that is why they so oppose each other, after all). Journalists, politicians, and concerned citizens all rightly pounced on this as inappropriate and wildly immoral, a "false equivalency." By doing so, they called him out on the grounds of comparison itself and isolated the cognitive mistake.

Comparison is a cognitive act that deals with difference, how the mind understands and frames in language an encounter with otherness. *Comparison is the cognitive negotiation of sameness and difference in a set of data.* Generally speaking, bad or inappropriate acts of comparison of other people are those that overemphasize either sameness (thereby eliding the real and important differences) or difference (thereby eliding or erasing the shared humanity). Good or appropriate acts of comparison of other people are those that balance sameness and difference, acknowledging both as important and refusing to deny one for the sake of the other. Social injustice derives from bad comparisons, as we saw in Charlottesville. Social justice relies on good comparisons: People who are different who have a right to be treated the same. Cognitively speaking, good comparison *is* justice.

Ultraconservative ideologues and their followers in the United States attempt to deny the integrity and importance of the differences of others and so subsume them into their own understanding of universal human sameness, which inevitably takes some form of whiteness or Protestantized Christianity. Intellectuals on the far Left make the opposite comparative mistake. They overemphasize human difference (usually coded as racial, gendered, or sexualized) to the point where all human sameness is relegated to the margins or demonized as "colonizing," "imperial," "hegemonic." They may do this for just moral reasons, but the results are nevertheless less than ideal: an endless fracturing of the liberal community into this and that political identity, which then fractures again, and again, and again, until there is no shared and universal *human* basis from which to act effectively and robustly in the political realm.

Perhaps so few people can balance sameness and difference in just and effective comparative acts because they have not been taught to do so. The cognitive structure and ethics of comparison, after all, do not appear anywhere in our school curricula. Perhaps it is so rare because it implies a moral and spiritual revolution that we have not yet gone through.

### 3. "The Religion of No Religion"

Nowhere is the failure to compare justly more apparent than with the difficult subject of religion. It is here that I think many well-meaning intellectuals and liberal politicians stumble and fall. Most fundamentally, I believe that liberal politicians and intellectuals have failed on the political or public level because they have either dismissed or not taken seriously enough the spiritual yearnings of their fellow human beings. They have not engaged, *really* engaged, religion. They have, accordingly, operated with a grossly inadequate anthropology—that is, with a model of the human that is not broad and deep enough.

Human nature is irrepressibly spiritual in nature in the simple sense that human beings will generally want to orient themselves to some larger whole or purpose. I would go so far as to suggest that it is quite correct to do so, since as individuals they *are* parts of a larger whole (the entire universe). They will do this in many ways, including through the stories, communities, rituals, and art forms of religious institutions. To the extent that any political platform attempts to deny, repress, or even downplay these spiritual impulses, it will inevitably fail.

The key is to affirm this shared spiritual nature without confusing it with any particular historical instantiation of that nature—that is, with any particular religion. The key is to affirm our sameness and our differences at the same time. The key is comparison.

The third way explored in these pages does not flinch when it comes to religion, and it knows how to compare them. It understands all religious phenomena as actualizations of a human potential that can never be exhausted or fully captured in *any* culture or *any* religion. It understands that we *need* culture and language and (often) religion, just as the artist needs paint, canvas, and brush. Religions are one way we paint ourselves, sculpt ourselves, author ourselves. The problem arises when the artists mistake their art forms for absolute blueprints or final descriptions.

Religious revelations, however, *do* claim absoluteness, finality. Within the history of Western monotheism, God speaks in booms and bolts and writes on rock that his (and, yes, it's always a he) is the *only* way, that he is the only God. *That's* the problem. We should resist and deny that "God," not because we know that there is no God (we do not), but because that "God" is not really God. He's just *us* (as in "us men") in another intolerant and violent form.

Much of this comes down to how we orient ourselves to history and whether we do or do not possess a robust historical consciousness that mourns as well as memorializes. We are suffering through a gigantic anachronism today. One can and should use the past as a *resource*, even be inspired by its various positive intellectual, artistic, spiritual, and cultural treasures, but one cannot and should not try to live in the past.

All historical representations, including and especially religious ones, encode not just religious or philosophical possibilities but past social practices and moral values, many of which are profoundly problematic, if not blatantly ignorant and frankly dangerous in the light of our modern knowledge and social circumstances. Most of our problems with religion today are a function of this massive mismatch: People are worshipping and trying to live out a set of scriptural texts that were written thousands of years ago in cultures none of us would want to live in now. *All* scriptural texts are fundamentally flawed, and they are killing us right now, literally. Every terrorist bomb shows this, but so does every suicide of every gay man tortured by a misinformed and misguided moral conscience, as does every man who thinks that he owns his wife and children like a farmer owns his field and its fruit.

These last attitudes reflect in a particularly acute and intimate way the massive social-moral anachronisms that we are suffering and the subsequent forms of ignorance they produce and support. Widely shared around the world by many, if not most, religions, such gender attitudes are carried in the metaphors of the ancient world, in which the essence of every person is believed to be in the "seed," which is in turn "sown" in the "fertile" or "infertile" "field" of the woman.

We know through modern genetics that a man and a woman each contribute twenty-three chromosomes (and that men can be infertile, too, something that the ancient texts could not imagine, given their agricultural metaphors of seed and soil). What would our moral values look like around gender and sexuality if we adopted this perfect genetic symmetry into our religious understandings?

Modern secularisms and the various spiritualities to which they have given birth do not generally make these anachronistic mistakes. They do not worship or honor the past in the same way that traditional religions do. Traditional religious faith sacralizes some past event or set of events to which believers are to stay faithful or believe in. Those events are ritualized, institutionalized, and mythologized. The faithful tell the same stories or myths over and over again and reenact them in the theater of ritual, festival, architecture, and institution. The result is a religious world rooted in the past and almost always unrealizable in the present. The inevitable results are guilt, anxiety, and violence.

Until, that is, one realizes that one can still appreciate and learn from past revelations and religious experience without feeling bound to their social, political, and moral expressions. One can turn around and look toward the future for more revelation and more religious experience.

Liberal American Christians sometimes say that God never uses periods when he speaks; he uses only commas. This clever grammatical dictum captures something important, Many in that wide swath of American culture that identifies itself as "spiritual but not religious," too, free themselves from such anachronisms. This demographic is comprised of individuals under thirty, but also of older groups, including and especially professional scientists.[3]

The roots of the contemporary spiritual orientation lie in the New Age movement of the 1980s and 1990s, the counterculture of the 1960s and 1970s, the human potential movement that originated in the 1950s and 1960s, and, further back still, in classical American figures like the writer Ralph Waldo Emerson and the erotic poet Walt Whitman, who recognized very early on that cosmic states of consciousness could best supply the ontological foundation for the political projects of radical democracy and social equality.

It was Whitman, often called the first and greatest of American poets, who first used the word *spirituality* in this modern sense (in his 1871 book, *Democratic Vistas*). He meant a mystical worldview that grounds democracy, a radical comparative orientation that saw all religions and all scriptures as organic and diverse expressions of an unacknowledged cosmic nature—our own. Here is the great poet in his greatest poem, *Leaves of Grass*:

> We consider the bibles and religions divine—
> I do not say they are not divine,
> I say they have all grown out of you and
> may grow out of you still,
> It is not they who give the life—it is you
> who give the life;

Leaves are not more shed from the trees or trees
from the earth than they are shed out of you.[4]

Note the flip here. We are not reducible to religions and their scriptures. They are reducible to us. Whitman and his followers considered this long poem and its teachings to be a new American Bible.[5] The first edition was published in 1855. The deep historical roots are older still. My first Ph.D. student was Hae Young Seong, a young man from South Korea who spontaneously awakened into an utterly transcendent, deeply blissful, orgasmic form of cosmic consciousness in a high school classroom when he was just sixteen. Years later, he would write his dissertation on Plotinus, in whose experiential descriptions of the One he recognized a more developed and mature form of his own teenage awakening, which at the time of the event he could not understand. Seong now reads that initial not knowing as a kind of wise respect for his own free will and as an implicit calling to pursue his own understanding and self-integration. He daringly writes of "IT" as the "ground of All Being" and calmly insists that this IT was the ultimate knowledge that a human being can attain: IT is where we are all from and where we are all headed.

These are *classical* claims in the comparative mystical literature, into which Seong's youthful awakening fits seamlessly. Seong's dissertation would make sense of his own earlier awakening and argue that Plotinus is the ancient father of the present spiritual but not religious orientation, tracing the orientation back almost two thousand years and locating it in the heart of the Western philosophical tradition, in one of the greatest mystical intellectuals of all time.[6]

Sadly, this is grossly misunderstood, even demeaned, among writers on both the Left and the Right. Left-leaning intellectuals tend to see little more than narcissism, intellectual fuzziness, and

"New Age woo-woo."[7] Conservative or fundamentalist religious leaders see a refusal to commit and a dangerous slide into secularism and relativism. Neither camp will own up to its own blind spots: a hyperrationalistic materialism and historical ignorance on the Left, a hyperliteralism and ethnocentrism on the Right.

I do not wish to romanticize or idealize the spiritual but not religious. The language of being "spiritual but not religious" remains more of an intuition, moral protest, or temporary placeholder than it is any kind of articulated moral vision or worldview. Most individuals who use this language, I'll venture, have little knowledge of any of this history or sophistication. But this hardly means that historical memory, philosophical rigor, and moral clarity cannot be articulated. A poorly articulated idea is not the same thing as a poor idea.

I wrote a history of the human potential movement in 2007, and years later, I agreed to serve on the board of the institute that originated the same movement, the Esalen Institute, in Big Sur, California.[8] I now serve as chair of that board. When I speak to the Esalen community, I always try to remind them that the movement is called the human *potential* movement and not the human *accomplishment* movement. I remind them that the movement does not look to the past (even the countercultural past) for some absolute revelation or answer, but to the future and to what Aldous Huxley called the "potentialities" of human nature for its developing and always tentative answers.

Huxley was yet another flipped intellectual; in his case, it took mescaline. He was fluent in scientific forms of thinking but also committed to the anomalous and mystical dimensions of human nature. He wrote about comparative mystical literature, famously introduced the word *psychedelic* into our vocabulary (through a friendship with a psychiatrist, who really came up with the word, Humphry Osmond), and was close friends with Eileen Garrett, one of the most eloquent and accomplished psychics of her time. Huxley, moreover, possessed a legendary scientific pedigree. His grandfather was none other than "Darwin's bulldog," T. H. Huxley, and his brother was the famous evolutionary biologist and scientific diplomat Julian Huxley. For these double humanist-scientific reasons, it is in the writer's notion of "human potential" that I find real hope and promise.

Stanford professor of comparative religion Frederic Spiegelberg inspired both of the founders of Esalen: Michael Murphy and Richard Price.[9] It was Spiegelberg, trained in medieval mystical theology and chased out of his native Germany during the Nazi oppression, who forged the notion of what he called the "religion of no religion."

Such a religion of no religion is not antireligion or simple secularism. It is inflected by the religious past and recognizes the cosmic truth of some forms of mystical experience (Spiegelberg had been flipped by his own mystical experience in a wheat field in 1917). But it also rejects identification with any particular historical, doctrinal, and institutional embodiment of those experiences. It sees these not as "choices," but as plural "expressions" of a shared human spirit. Significantly, Spiegelberg stressed that the religion of no religion is ultimately creative of new religious forms. This "new" spirituality, then, possesses its own paradoxical logic, which, once understood, is rigorously rational and firmly rooted in some of the deepest and most important currents of global mystical thought.

### 4. "Cosmic Humanism"

Much of this spiritual revolution is already implied in the most basic values and methods of the humanities, even if this has seldom, if ever, been spelled out and has, in fact, been actively resisted in the academy for about four decades now on the grounds that any invocation of "humanity" must privilege some specific ethnic or racial model of humanity, usually a European and white one.

Such critiques are well aimed, morally just, and basically correct with respect to many past visions of the human. But again, a badly used idea is not the same thing as a bad idea, and no idea of justice can even make sense without some underlying basis of universality or equality. As I listen to public lecture after public lecture and read dissertation after dissertation on this or that historical period, cultural figure, religious system, or piece of literature at my home university, it is patently obvious to me that there is a deep substructure or superstructure at work in the humanities, one that works from a logic of the whole to a local logic of the parts, much as we saw with Alex Wendt's quantum social science. *The human whole is prior. The human whole is fundamental.*

That is why *any* human expression from *any* historical period from *any* culture can be studied in the humanities with the same methods and the same questions. If human nature were not one, it would simply not be possible. If human nature were not multiple, it would not be necessary. It is this balancing of unity and diversity, sameness and difference that is the most basic challenge, and the most basic gift, of the humanities.

The problem here is that all of our national, cultural, and religious identities work from the opposite logic. They privilege the parts over the whole and deny the whole in various subtle and not so subtle ways. Fundamentalist Christians do not work from the superlogic of a shared humanity upon which is built particular, local, and historically relative Christian identities (all different, of course). Neither do fundamentalist Muslims, Catholics, Jews,

Hindus, Buddhists, and so on. Each of them presumes a specific religious universal nature (its own) and argues, implicitly or explicitly, that our shared humanity is either secondary or unimportant compared to whatever particular religious identity they happen to be privileging. In short, the conservative religious mind works in the opposite direction than that of the liberal humanities.

The devastating political result of this religious upside downness is the social production of a set of religious identities that assume that they are ultimate and universal, when, in fact, they are all historically constructed, relative, and temporary. These socially produced identities then become the basis upon which our Newtonian billiard-ball communities, nation-states, and religions collide, bounce off one another's illusions, and ultimately go nowhere, or, worse yet, result in open conflict and collision (being imagined billiard balls and all).

During the 2016 presidential race, after he was named the vice presidential candidate for the Republican Party, we heard Michael Pence say these lines: "I am a Christian, a conservative, and a Republican—in that order."[10] Note that he left out what he was *before* all of these things; that he was a human being. Versions of this prioritizing of a religious identity over a shared humanity could be easily found in fundamentalist leaders around the world, each of whom is arguing for similar forms of privilege and dominance for his or her religious community and thus doing damage to his or her larger community and shared species.

What if we simply stopped doing this? What if we stopped identifying humanity with *any* religion, nation-state, or ethnicity? We do not need to keep playing this game. We can be someone else.

I recognize how utopian that all sounds. So allow me to propose a more humble and realizable beginning. What if we all accepted and embraced our own specific cultural or religious identities, yes, but stopped seeing any of these as ultimate, and started seeing them all as expressions of our shared humanity? Herein lies the potential politics of the flip. We are not our thoughts. We are not even our beliefs. We are first and foremost conscious and cosmic, which is to say we are human.

How do we deny humanity to someone simply because he or she has a different color of skin, or holds a different set of beliefs in her or his head? How is this even *imaginable* within such a cosmic humanism? The simple answer is that it isn't. And once we recognize this, once we become human again, and truly cosmic, we will stop enacting these petty bigotries in our politics and societies. They will become unthinkable, impossible.

## 5. "Deep, Dark Ecology"

It is not clear how much time we have to render what is now politically powerful morally unimaginable. Though many of our citizens are still in denial, we are in the midst of a global climate crisis and another great extinction event, and we are the primary cause of both. This is where reflexivity—stepping out of one's deepest held thoughts and beliefs—becomes globally urgent. If we do not recognize that we ourselves and *our most cherished beliefs are themselves the problems* (and by "we," I refer mostly, but not exclusively, to those of us who live in developed economies and so have burned the vast majority of the carbon that has warmed and polluted the Earth), how can we effectively address such a global crisis? Indeed, how will we even *think* it if our thoughts are simply reexpressions of the very thoughts that created the crisis in the first place?

Amitav Ghosh has warned us that the unquestioned truths of our now outdated political and economic beliefs (like sovereign "individuals," "private property," "nations," infinite economic "growth" without consequence, even "moral uplift") are powerless to address these ecological crises and, indeed, only exacerbate and make them immeasurably worse.[11] Does the supposed "competition" of free-market capitalism, driven, of course, by self-interest and the accumulation of wealth in tiny, tiny pockets of the species and with little or no eye to the larger environment, make any sense at all in the Anthropocene? How about the "freedom" to build coal plants and purchase endless automobiles that pollute everyone's atmosphere and warm the planet to who knows what effect? All of these acts may seem natural and justifiable in a Newtonian imagination, where we are individuals seeking wealth with no real connection to or responsibility for one another and the environment. They make no sense at all in the new quantum real, where we are intimate expressions of a shared single cosmos.

This is where many traditional forms of humanism clearly fall and finally fail. Humanism has been justly criticized as anthropocentric—that is, as too centered on the human species at the expense of other life-forms and deeper ecological networks. A truly cosmic humanism, however, extends through the natural order into the universe. It humanizes all forms of consciousness, wherever they are found. This is what makes it a species of a deep and dark ecology. Both deep ecology and dark ecology possess specific literatures and nuances, even if they have much in common.[12]

Deep ecology is a broad spiritual and philosophical movement that can be traced back to the Norwegian intellectual and activist Arne Næss, who was a trained philosopher and a student of Mahatma Gandhi. Its basic claim is that we are an intimate part of a larger ecobody or ecosystem and so should care for the ecosystem not as a collection of dead objects or neutral resources "out there," but *as our own larger body.*

Crucially, there is no external deity here issuing commands or moral systems. Such a worldview sees a deep ecological life as an expression of self-defense and self-care, since we *are* the natural world. This worldview is a perfect expression of the kinds of humanistic-scientific fusions that I am proposing here, since its basic vision of the human as an expression of the natural world sits seamlessly within a set of cosmological, biological, and physiological facts. Science is not sufficient here, though, since science alone cannot produce or articulate such moral values. The moral vision of deep ecology needs the sciences for its facts and understanding of the natural world and human nature, but the sciences need the philosophy and the spiritual vision to articulate such an ethical vision.

In a similar spirit, the scholar of religion and nature Bron Taylor has written of what he calls "dark green religion." Taylor defines the latter as any spiritual worldview that understands and treats the natural world as sacred or ultimately meaningful *in its own right*, and not as some separate "creation" of an external god.

This is where a renewed panpsychism might really help us. Taylor has reminded us of the disturbing colonial and ethnocentric connotations of the term *animism*, and how Western thinkers have used the category to argue that worldviews that emphasize the fundamental deadness of reality are "advanced," whereas worldviews that emphasize the fundamental aliveness of reality are "primitive."

Not so long ago, Amazonian cultures were considered ridiculous for holding that plants—especially psychoactive ones that engage human beings in countless acts of healing, guidance, and revelation—possess agency, intelligence, and consciousness.[13]

Now we are told by cutting-edge botanists that plants possess all kinds of sophisticated forms of intelligence; that they interact with their environments in intentional and agential ways, if within a different or "slower" temporal dimension; that they send electrical and chemical signals between their cells like animals; and that there may even be something we might call "chlorophyllic sentience."[14] They express what looks like brainy behavior without having any brains.[15] When Amazonian shamans code insights in elaborate mythical or visionary ways, it is "magical thinking" or "animism," but when botanists code in technical or purely descriptive ways, it is "science."

This is not just a moral complaint, it is a philosophical and scientific point. We would benefit from listening to the history of religions with a very careful ear, for human beings have been *experiencing* extremely advanced forms of knowledge about the natural world for millennia.

Timothy Morton's "dark ecology" is especially relevant here.[16] The expression names a very special "loop structure" of ecological thinking that he calls "ecognostic," which he defines as "a knowing that knows itself."[17] He finds this direct knowing in ancient gnostic symbolism, particularly in the central gnostic symbol of the *ouroboros*, or the snake biting its own tail. This loop structure was dramatically displayed much more recently in one of Morton's favorite films, *Blade Runner*, the classic 1982 sci-fi dark noir film directed by Ridley Scott and based on the Philip K. Dick 1968 novel, *Do Androids Dream of Electric Sheep? Blade Runner* features a protagonist named Rick Deckard, who spends the entire film hunting replicants, or androids, only to suspect, in Morton's reading now, that he himself may be a replicant.

It is this uncanny realization that defines the Anthropocene for Morton. We are all living in a Philip K. Dick reality now, a world in which we are the main actors on the global stage, and everything we think we are doing to and with "nature," we are actually doing to and with ourselves, since we *are* nature become conscious of itself. We are hunting, farming, polluting, warming, and destroying ourselves.

Morton explores this argument through the fundamental insight that all of life is interconnected and constitutes a single plural whole, and that there really is no such thing as "nature," that is, a natural world that is somehow set apart or other than us.[18] He intuits a new humankind for the future, at one with all there is, in solidarity "with all non-human people," and very much in the process of cocreating or "evolving" itself.[19] Dark ecology, he writes, "is how we find ourselves in a story we have written, and the next part is becoming conscious authors of the story we are writing."[20] Morton detects this future humankind in particular genres of art, particularly science fiction, for "art is thought from the future"—that is, "[t]hought we cannot presently think at the present."[21]

Religion itself can be uncanny or looplike in precisely this way. As the French sociologist Émile Durkheim and his followers have put it: Religion is society worshipping itself. Or as Freud had it in *The Future of an Illusion*, all the gods and every God is a projection of the human, and an immature, if not actually infantile, one at that. So here is my proposed flip: If all gods are in fact projections of us, then there is something about us, which we cannot yet own or admit, that is godlike.

Such a realization of being caught in a story we ourselves are writing as unconscious gods can be terrifying, depressing, even nihilistic. Morton attempts to move us from the dark depression to the sweet gnosis of directly knowing who and what we already are. And what we already are is truly fantastic: 5-D or

"X-beings" endowed with spectral superpowers that give witness to the future human, a massive transdimensional hyperobject awakening to its own actual condition, connected to everything there is, and literally and truly cosmic.[22]

Not surprisingly, indeed predictably by now, such sci-fi language is the result of Morton himself being flipped within eerie paranormal moments of his own:

> So, the 5-D Them, the superbeings, are not super in the sense of having transcended time and space or having achieved ultimate mastery; nor is being your own author a question of mastery, but a deeply confusing matter of self-haunting, as anyone who has had a paranormal experience will affirm. They are deeply ambiguous. Am I having something real or is this an illusion or delusion? What type of illusion? What form of real? Am I in someone else's story or am I writing my own?[23]

There is the flip as reflexivity, as the quintessential insight and key skill of the humanities as prophetic witness.

### How (Not) to Make Fundamentalism Stronger

Over a half century ago now, C. P. Snow (1905–1980), an English chemist and novelist, argued in a slim little book entitled *The Two Cultures and the Scientific Revolution* that we have two separate and largely incommunicable intellectual cultures in the West: a humanistic or literary one, and a scientific or technological one. He saw this division and our inability to speak and think across it as a fundamental obstacle to our adequately addressing our deepest and most intractable problems. In the last six decades, we have seen an increasing rift between the two intellectual cultures and a more and more troubling silencing and defunding of the humanistic side.

But more ominous is *a second split*: that between the educated and the uneducated, with the result being a profound anti-intellectualism or rejection of all forms of professional knowledge, be they humanistic *or* scientific. Hence the shockingly common denial of evolution and climate change and the demonization of liberalism in all its forms. This split is another reason why scientists need humanists, and why humanists need scientists. We are now in the cultural trenches together.

The flip is poised directly *against* any and all simple returns to some past religious authority. As I have explained, a flipped individual may turn to previous religious systems for inspiration and guidance, or belong to a religious community, but she or he will not be slavishly bound to any of this, as she or he will understand the religions as humanly constructed responses to some earlier flips, not absolute truths to literalize, "believe," and worship.

Contrary to what is often assumed, what we call "fundamentalism" is an eminently modern phenomenon. Most historians of American religion trace its definitive beginnings back to a series of pamphlets called "The Fundamentals," which were issued in the second decade of the last century. The movement was complex, but it was poised against two broad forms of professional knowledge and expertise: the academic study of religion (particularly biblical criticism or the historical study of the Bible, which had shown again and again how the biblical texts were constructed by human communities and political processes) and modern science (particularly evolutionary biology, which had shown how human beings had evolved from earlier primates). The modern mind-set of fundamentalism originated in a rejection of *both* the humanities *and* the sciences.

Research on fundamentalist movements around the globe has shown that there have been two especially common career paths from which fundamentalist leaders have often emerged: engineering and computer science. I suspect cognitive and educational reasons for this. These individuals tend to think in very literal and literally binary terms. Computer codes and blueprints for bridges, after all, cannot contain any ambiguity, much less open paradox.

But that is not how religious writings and religious experiences work. Ambiguity and ambivalence, even open paradox, lie at their very heart. Such forms of reflexivity and ambiguity also lie at the heart of the humanities, which is why fundamentalist leaders hate the humanities and humanists.

We need to understand that fundamentalism is also a symptom or a cancer, an out-of-control immunological response of the social body responding to the toxic environment of nihilism and meaninglessness that defines so much of modern Western society and to which scientists and humanists have contributed more than their fair share.

Sometimes one can best understand an idea by looking at those who reject it. Materialist debunkers and religious fundamentalists sound remarkably alike in their rejections of the kinds of extraordinary experiences I have engaged and analyzed here. This is because they have cocreated one another. Accordingly, there is no way out of our present impasse in the exclusive terms of either camp. They only can reinforce and strengthen each other.

The third path proposed here is a cosmic humanism that is deeply religious without being religious, a human expression of awe and beauty before a living conscious cosmos that transcends any and all human efforts to comprehend, much less explain, it, be these "religious," "scientific," or some future form of mind and knowledge that we can barely imagine at the moment.

## On Being Right

But am I right about any of this?

The phrase "being right" encodes a neuroanatomy that we are learning more and more about with each passing decade. Our cognitions, our languages, and our metaphors are expressions of our bodies and their kinesthetic orientation and movement through space and time. All our thoughts are embodied thoughts.

The right side of the body is controlled by the left hemisphere of the brain, where language, linear and mechanistic forms of thought, mathematics, and—tellingly—the social ego all primarily (but not exclusively) dwell. The left side of the body is controlled by the right hemisphere of the brain, where our sense of wholism, comparativism, symbolism, and intuitions primarily (but not exclusively) lie.

Most cognitive, linguistic, emotional, and sensory functions, however, are "global"—that is, distributed across both hemispheres. Take, for example, the extremely complex neurological correlates of language. Whereas the left hemisphere is involved in generating the meaning of individual words, the right hemisphere is in charge of grasping more wholistic meanings, such as narrative and humor, and correlating feeling or emotion to content.[24] Even when the tasks are distributed, the basic lateralization of the brain is maintained. Moreover, however much we want to emphasize the global distribution of brain function, the two brain hemispheres function relatively separately as well.

I take it as a given that all the "right" and "left" symbolism of Western culture is a function of this same neuroanatomy and, more particularly, of the left hemisphere's tendency to demean and deny the right hemisphere. Jesus sits on the right side of God, not on the left. The Latin for "left" is *sinister*. When my father tried to write with his left hand in grade school, the nuns would strike it with a ruler. The "hard" mechanistic sciences now trump the "soft" intuitive and artistic humanities. "The Right" now demonizes "the Left."

• • •

Jill Bolte Taylor has spoken to precisely this neuroanatomical split and its intellectual, moral, emotional, and political dimensions and, by so doing, became a genuine prophet of a new cosmic humanism. Her TED Talk "My Stroke of Insight," later published as a book with the same title, now stands at over twenty million views and is listed as one of the twenty most viewed TED Talks of all time.[25]

On the morning of December 10, 1996, Taylor had a massive stroke that exploded in the left hemisphere of her brain. Since she was working at the time as a neuroanatomist, she was well aware of what was happening to her during the stroke itself. She humorously describes how "cool" it was for a brain scientist to experience a stroke from the inside.

But it was also potentially deadly. As her rational, logical, and language-oriented left brain shut down during the stroke, she entered what she affectionately calls "La-La Land" and "Nirvana," her alternating terms for a state of consciousness defined by an oceanic state of eternity, living energy, omniscience, tranquillity beyond measure, blessedness, and an utter conviction of being one with the entire universe.

Taylor walks a tightrope, as her language carefully balances and sets into conversation the  mystical and scientific, holistic and reductive, or Platonic and Aristotelian, modes of thought (for our purposes anyway, Plato sits in the right hemisphere of the brain, whereas Aristotle sits in the left).

Although clearly Taylor is "of two minds" (literally and metaphorically) and would affirm both ways of thinking, she uses explicit mystical language in *My Stroke of Insight*. Her story is about the trauma *through which* her insight came, or, we might say, was let in.

She uses plenty of reductive language in the book, but its ends are, instead, "panpsychic" and "emergentist." She thus relates different forms of consciousness that arise from the collective cooperation of trillions of living intelligent cells, or she describes the deep inner peace she felt as being the "neurological circuitry located in our right brain."[26] "I believe the experience of Nirvana exists in the consciousness of our right hemisphere, and that at any moment, we can choose to hook into that part of our brain."[27]

Is Nirvana nothing more than a particular expression of the right hemisphere of the material brain? Such rational language alternates throughout the text with openly mystical or spiritual language. As the stroke sets in, Taylor's initial sense of things became "bizarre, as if my conscious mind was suspended somewhere between my normal reality and some esoteric space." She had become "trapped inside a meditation that I could neither stop nor escape."[28] In another place, she describes how she was "suspended between two worlds, caught between two perfectly opposite planes of reality."[29] The accent, it seems, does not go here or there, but on both. Hence her answer to Carl Sagan. "I love knowing that I am simultaneously (depending on which hemisphere you ask) as big as the universe and yet merely a heap of star dust."[30]

She describes flying into "a void of higher cognition" and how she "soared into an all-knowingness."[31] Her brain-body is "a portal through which the energy of who I am can be beamed into a three-dimensional external space," and she sees this body-portal as "a marvelous temporary home." She expresses astonishment at how she could have spent so many years unaware of this, never really understanding "that I was just visiting there."[32]

Virtually every word and phrase that Taylor uses ("all-knowingness," "void," "Nirvana," the "visit," "omniscience," "eternal," and so on) possessees a rich history in mystical literature. Some of the comparative resonances are striking.

She constantly invokes the language of "energy" to describe not just human beings but the way the right brain perceives the world. "Because everything around us . . . is composed of spinning and vibrating atomic particles, you and I are literally swimming in a turbulent sea of electromagnetic fields. We are part of it. We are enveloped within it, and through our sensory apparatus we experience *what is*."[33] Or, as she feels herself dying, she "felt the enormousness of my energy lift. My body fell limp, and my consciousness rose to a slower vibration."[34] She explains how she came to understand herself as "a being of light radiating life into the world,"[35] that she is part of "an eternal flow of energy and molecules from which I cannot be separated," and how her right mind knows that "the essence of my being has eternal life," in that when she dies, "my energy will merely absorb back into the tranquil sea of euphoria."[36]

Much as Alexander Wendt believes that human beings are literally "walking wave functions," for Taylor words like *energy, vibration,* and *radiating* are not metaphors. Energy seems to be conscious and alive. We can know this directly, because "[o]ur right brain is capable of detecting energy beyond the limitations of our left mind because of the way it is designed, . . . we are energy beings designed to perceive and translate energy into neural code."[37]

Along with the physics of energy, Taylor invokes evolutionary biology and insists on the role that we can yet play in our own evolution, pointing out the modern human brain is quite different from the human brain of even a few thousand years ago. Language use in particular has physiologically "altered our brains' anatomical structure and cellular networks."[38] Toward the end of the book, she picks up on the metaphor of "tending the garden" of the brain to suggest something of our own future evolution, how we can tend to the garden of neurons, nourishing some roots and branches and pruning back others. She is not being metaphorical with such evolutionary language.

And so we come to Taylor's moral and political vision. The left and right minds do not just perceive the world differently. They have entirely different *values*. Taylor locates the separate and socially conflictive ego in the left hemisphere of the brain and sees a shared cosmic humanity beaming into and through the right hemisphere. Accordingly, the primary moral expressions of the right brain are a profound sense of the equality of all human beings, a felt sense of the whole human family, a lack of arrogance or aggression toward others, and, above all, a profound sense of compassion. The right mind does not perceive or heed artificial differences, like boundaries, territories, race, or religion.[39]

Taylor also describes how when she was identified with the consciousness of the right brain, it was impossible for her to feel either physical or emotional loss. She could not suffer, not because she was immune to loss or physical pain, but because she was no longer an individual separate from everything else. She *was* everything else.[40]

Because we do not really understand the different functions

of the two hemispheres, and because we are presently more or less cut off from the wisdom and eternal being of the right brain, we are at the mercy of our own mental and emotional states. We identify completely with them, as if they were us. Accordingly, we are also at the mercy of external threats, such as advertising and political manipulation.

At the climax of her TED Talk, Taylor pleads with her scientific colleagues to step to the right, to cease identifying only with their left brains, so that we can all begin to make the world a better place. One can sense an awkward, stunned silence in the audience.

# THE COSMIC HUMAN

*When everything is human,*
*the human is a very different thing.*

—EDUARDO VIVEIROS DE CASTRO, *Cannibal Metaphysics*

For reasons that we have not begun to realize, the universe, the elements, the basic laws of physics that shape and guide everything there is eventually result in conscious life-forms we call "human beings" or "people," as if that consciousness were already somehow encoded in the cosmos. Once we make this panpsychist move in which everything is "conscious" or "human," everything changes, including and especially what we think of as the human. The cosmos is not just human. The human is also cosmic.

• • •

I am certain that many will consider my emphasis in this book on extreme religious experiences and anomalous states of consciousness an eccentric means to argue for the future centrality of the humanities, much less the irreducible nature of consciousness. I make no apologies. We must be bolder. We must proceed through an intentional and systematic ontological shock if we are ever going to arrive at the future of knowledge.[1]

We should remember that early science arose in what were essentially private clubs and confidential social spaces outside the university, the church, or any other official structures, mostly because the originators knew perfectly well just how heretical and incompatible their new knowledge was vis-à-vis the reigning forms of knowledge. If the past is any measure, the future of knowledge, too, may emerge from the margins.

I think the future form of knowledge will be, like the human brain itself, *doubled but equal*—that is, reciprocal. Plato will have as much to say as Aristotle. The quantum physical will be seen to complement the classical material, as the right hemisphere does the left hemisphere. Holism, comparativism, and intuition will be as valued as linear logic, mathematics, and reductive analysis. We will realize that there is no study of objects without subjects, or vice versa. Educationally speaking, the humanities will not be the poor cousin of the sciences.

Our deepest questions about ourselves (as parts) and the cosmos (the whole) will never be answered without an engagement with both forms of knowledge. And such ultimate questions will probably require forms of knowledge that we have not yet even imagined.

Wolfgang Pauli, the quantum physicist, had it just right when he wrote these lines:

> Contrary to the strict division of the activity of the human spirit into separate departments—a division prevailing since the nineteenth century—I consider the ambition of overcoming opposites, including also a synthesis embracing both rational understanding and the mystical experience of unity, to be

the mythos, spoken and unspoken, of our present day and age.[2]

It is time to tell this new story or mythos now.

Quantum physics is a "flipped science," one in which consciousness is no longer understood as an epiphenomenon, but as fundamental to the very nature of nature. If consciousness is fundamental here, it stands to reason that it will someday become so in evolutionary biology and cosmology, which depend entirely on the quantum physical field as their substratum and base.

How could it be otherwise? Dramatic and life-changing flips aside, it is difficult to read the various philosophers and scientists on panpsychism, dual-aspect monism, quantum mind, cosmopsychism, and idealism and not wonder about the limitations of neo-Darwinian materialism—that is, the present assumption that the cosmic, evolutionary, and genetic processes are purely random and without agency, consciousness, or intention. If the "inside" of matter is mind, as all of these new models suggest in different ways, how can neo-Darwinian materialism be the full truth of things? The answer is simple: It cannot be. None of this need involve any form of intelligent design, much less any biblical literalism or creationism (with an external deity "creating" or "causing" a material world). There are other sophisticated models on the table, as the iconic case of Einstein clearly shows, and as we have seen over and over in these pages.

We can turn things around. As the evolutionary biologists realized some time ago, social and cultural evolution and their effects (things like tool use, language, social organization, and agriculture, and now the industrial revolution, the Pill, the silicon revolution, genetic engineering, and climate change) have taken over, and dramatically quickened evolutionary processes that were once determined almost entirely by biology and adaptive pressures. So, too, have intellectual, moral, and spiritual disciplines quickened the pace: practices like education, reflexivity, and the flip as contemplation—that is, the practice of stepping away from one's thoughts, emotions, and beliefs, even one's sense of a social self, and becoming more and more conscious of consciousness itself.

# Acknowledgments

M any people helped me with this book, first and foremost
Erika Goldman, without whom there would be no book.
It was Erika who first read my essay on this same set of ideas in
*The Chronicle of Higher Education*, contacted my literary agent,
Anne Borchardt, stuck with me over the years as I stumbled along
(and took too long), and then made it a much better book with
her careful editing. Anne also deserves much credit, because she
encouraged me to take up the project when I was feeling overwhelmed by other projects, as did Jean Tamarin, my editor at *The
Chronicle of Higher Education*, who encouraged and published
that first exploratory piece. So, Erika, Anne, and Jean, thank you.

The book is also deeply indebted to all of the scientists, philoso-
phers of mind, historians of science, social scientists, and human-
ist scholars with whom I have interacted around this text, through
established collegial networks, correspondence, and at the Esalen
Institute, in Big Sur, California, where I work closely with Michael
Murphy to host symposia on these same science-and-spirit topics.
Here I would especially like to thank Eben Alexander III, Harald
Atmanspacher, Renaud Evrard, Bruce Greyson, Menas Kafatos,
Bernardo Kastrup, Edward Kelly, Tanya Luhrmann, Paul Marshall,
Timothy Morton, Steve Paulson, David Presti. Gustavo Rodrigues
Rocha, Mark Ryan, William Seager, Hae Young Seong, Andreas
Sommer, Alexander Wendt, and Marjorie Hines Woollacott.

I would also like to thank Sidney Harris and ScienceCartoons-Plus.com for permission to print the cartoon "I think you should be more . . ."

The book is very much an original piece of writing, but, like all writing, parts of it are indebted to or drawn from earlier writings in previous publications. I would like to personally thank these presses for their gracious support of this project in the form of permissions to reuse material: the University of Chicago Press, for permission to use select material from my recent memoir-manifesto *Secret Body: Erotic and Esoteric Currents in the History of Religions* (2017); Wiley-Blackwell, for permission to use select material from *Comparing Religions: Coming to Terms* (2014), in which I explored my understanding of the humanities as the study of consciousness coded in culture and my category of the super natural; and Bloomsbury Academic, for permission to use the material on Marjorie Hines Woollacott, which first appeared in a foreword entitled "Of Saints and Scholars," in *Hagiography and Religious Truth: Case Studies from the Abrahamic and Dharmic Traditions*, ed. Rico G. Monge, Kerry P. C. San Chirico, and Rachel J. Smith (2016).

I have done my best to trace and acknowledge these debts, and all others, in the endnotes. I apologize beforehand for anything I may have unknowingly and unintentionally missed. I confess to taking some moral (and metaphysical) comfort here in the basic premise of quantum physics that everything is finally one thing. If this is true, then our debts are not only endless and unfathomable, they are also everything.

# Endnotes

## Prologue: The Human Cosmos

1. Chapter 1 is an expanded version of an essay that I published a few years ago: "Visions of the Impossible: How 'Fantastic' Stories Unlock the Nature of Consciousness," *The Chronicle of Higher Education*, April 4, 2014; https://www.chronicle.com/article/ Embrace-the-Unexplained/145557.

2. I am indebted for my language of the "third way" to the American psychologist and philosopher William James, who, in turn, inherited it from an earlier British psychical research tradition, where it was sometimes expressed in its Latin form, *tertium quid*. For a recent book-length study of James's third way, see Krister Dylan Knapp, *William James: Psychical Research and the Challenge of Modernity* (Chapel Hill: University of North Carolina Press, 2017). Perhaps it is significant that one of the earliest occurrences of the Latin expression was in the fourth-century debates about Christology on how to relate the "divine" and "human" natures of Christ. Apollinaris argued that Christ was "neither human nor divine but some 'third thing'" (Knapp, 6). I take these as early debates about how to relate mind and matter within a flipped individual, here through the local historical prism of Christian theology.

## 1. Visions of the Impossible

1. *Autobiography of Mark Twain*, vol. 1, ed. Harriet Elinor Smith (Berkeley: University of California Press, 2010), 274–276.

2.  See "Mental Telegraphy" (1891) and "Mental Telegraphy Again" (1895), in Mark Twain, *Tales of Wonder*, ed. David Ketterer (Lincoln: University of Nebraska Press, 2003), 96–111.

3.  Janis Amatuzio, M.D., *Beyond Knowing: Mysteries & Messages of Death and Life from a Forensic Pathologist* (Novato, CA: New World Library, 2006), 84–85.

4.  For a brief history of the supernatural and its transformation into the preternatural and the paranormal, see my "Introduction: Reimagining the Supernatural in the Study of Religion," in Jeffrey J. Kripal, ed., *Super Religion* (Farmington Hills, MI: Macmillan Reference USA, 2016). For my discussion of Maxwell, I am indebted to Renaud Evrard and his essay "Joseph Maxwell: Jurist, Physician, Psychist (1858–1938)," *Psypioneer 5, no.* 1, (January 2009), 21–27.

5.  I first employed this expression in Jeffrey J. Kripal, with Ata Anzali, Andrea R. Jain, and Erin Prophet, *Comparing Religions: Coming to Terms* (Oxford: Wiley-Blackwell, 2014), ch. 4.

6.  Quoted and discussed in Andreas Sommer, "Crossing the Boundaries of Mind and Body: Psychical Research and the Origins of Modern Psychology" (Ph.D. diss. University of London, 2013), 22. The original text is Francis Bacon, "Of the Proficience and Advancement of Learning, Divine and Human" (1605).

7.  This is an extreme and telling case. See especially Betty Jo Teeter Dobbs, *The Janus Faces of Genius: The Role of Alchemy in Newton's Thought* (Cambridge: Cambridge University Press, 1991). Newton, it turns out, was not simply Newtonian.

8.  The notion of bimodal concepts is indebted to the recent work of Jason A. Josephson-Storm in *The Myth of Disenchantment: Magic, Modernity, and the Birth of the Human Sciences* (Chicago: University of Chicago Press, 2017).

9.  See Elizabeth G. Krohn and Jeffrey J. Kripal, *Changed in a Flash: One Woman's Near-Death Experience and Why a Scholar Thinks It Empowers Us All* (Berkeley: North Atlantic Books, 2018).

10. Ann Taves, "Building Blocks of Sacralities: A New Basis for Comparison Across Cultures and Religions," in *Handbook of Psychology of Religion and Spirituality*, 2d ed., ed. Raymond F. Paloutzian and Crystal Park (New York: Guilford Press, 2013). For a full demonstration of the method, see Ann Taves *Revelatory Events: Three Case Studies of the Emergence of New Spiritual Paths* (Princeton: Princeton University Press, 2016).

11. For a textual source, see Kary Mullis, *Dancing Naked in the Mind Field* (New York: Pantheon Books, 1998), 130–136.

12. For examples, see Phil Zuckerman, *Society Without God: What the Least Religious Nations Can Teach Us about Contentment* (New York: New York University Press, 2008), 45, 54–55, 78–79, 86–87, 92–93, 135, 145–148.

13. Ibid., 147–148. I describe a near-identical scene involving a mother and a small child in Kripal, *Comparing Religions,* 366. Mothers and children are emotionally, physically, even genetically entangled in profound ways, of course. Hence, I would suggest, the common telepathic or clairvoyant cognitions that spark and spike between them. Perhaps there is even a paranormal biology here, whatever that might mean.

14. Zuckerman, *Society Without God,* 146.

15. *Kant on Swedenborg: Dreams of a Spirit-Seer and Other Writings,* ed. Gregory R. Johnson, trans. Gregory R. Johnson and Glenn Alexander Magee (West Chester, PA: Swedenborg Foundation Publishers, 2002).

16. Freeman Dyson in Elizabeth Lloyd Mayer, Ph.D., *Extraordinary Knowing: Science, Skepticism, and the Inexplicable Powers of the Human Mind* (New York: Bantam, 2007), viii–ix.

17. I am perfectly aware that the analogy is imperfect, in that boiling water does not, in fact, reduce it to the two gases. Only other, more extreme processes will do this. Still, the point is clear enough: Ordinary water looks nothing like two gases fused together by invisible forces, even though that is exactly what it is.

18. See https://tif.ssrc.org/2017/12/11/supernatureculture/.

19. For a model collection of essays and a powerful introduction to the new materialisms, see Diana Coole and Samantha Frost, eds., *New Materialisms: Ontology, Agency, and Politics* (Durham: Duke University Press, 2010).

20. Thomas Nagel, *Mind and Cosmos: Why the Materialist Neo-Darwinian Conception of Nature Is Almost Certainly False* (New York: Oxford University Press, 2012).

21. I originally made this point in Kripal, *Comparing Religions*, 86–87.

22. Victoria Nelson, *The Secret Life of Puppets* (Cambridge: Harvard University Press, 2003); Wouter J. Hanegraaff, *Esotericism and the Academy: Rejected Knowledge in Western Culture* (Cambridge: University of Cambridge Press, 2013).

23. *Specimens of the Table Talk of the Late Samuel Taylor Coleridge*, vol. 1 (London: J. Murray, 1835), entry for July 2, 1830. I am indebted to Eben Alexander for this quote. He uses it in *The Map of Heaven: How Science, Religion, and Ordinary People Are Proving the Afterlife* (New York: Simon & Schuster, 2014), 1.

24. Nelson, *The Secret Life of Puppets*, 16.

25. Such a hypothesis does not answer the question of the precise relationship of mind and matter. The filter thesis is compatible with both dualism and idealism. I discuss some of the different options in chapter 3.

26. David Eagleman, *Incognito: The Secret Lives of the Brain* (New York: Pantheon, 2011), 221–222.

27. William James, *Human Immortality: Two Supposed Objections to the Doctrine* (Boston: Houghton, Mifflin, 1898), 15.

28. G. William Barnard, *Living Consciousness: The Metaphysical Vision of Henri Bergson* (Albany: State University of New York Press, 2011), 237.

29. Quoted in ibid. The passage is from Henri Bergson, *The Creative Mind*, trans. Mabelle L. Andison (New York: Philosophical Library, 1946), 69–70.

## 2. Flipped Scientists

1. Bruce Greyson, personal communication, related in Marjorie Woollacott, *Infinite Awareness: The Awakening of a Scientific Mind* (Lanham, MD: Rowman & Littlefield, 2015), 207.

2. For a fruitful start, consider the psychologist Charles Tart and his collection of eighty contemporary examples, which he posted on the website TASTE, for the Archives of Scientists' Transcendent Experiences. See www.issc-taste.org.

3. David E. Presti, *Foundational Concepts in Neuroscience: A Brain-Mind Odyssey* (New York: W. W. Norton, 2016), 202–203.

4. D. Millett, "Hans Berger: From Psychic Energy to the EEG," *Perspectives in Biology and Medicine* 44, no. 4 (2001), 522–42, quoted in Presti, *Foundational Concepts in Neuroscience*, 202–203.

5. I have relied here on the Wikipedia entry on Berger, which, in turn, relies on the biographical essay at: http://www.encyclopedia.com/people/history/historians-miscellaneous-biographies/hans-berger.

6. Acknowledging that the philosopher was elderly and suffering failing health, it is still necessary to observe that Ayer's comparative base here is laughably small: just two cases. We now have archival files on *thousands* of near-death experiences. We know that light forms are often encountered, but we also know that the colors of this light morph, like the bands of a rainbow. We could go on for

pages here about red lights, blue lights, pink lights, a strange kind of black light, white lights, and so on.

7. A. J. Ayer, "What I Saw When I Was Dead," *Sunday Telegraph*, August 28, 1988.

8. See especially Emily Williams Kelly's introduction and anthologization of Stevenson's essays and writing in *Science, the Self, and Survival After Death: Selected Writings of Ian Stevenson* (Lanham, MD: Rowman & Littlefield, 2013). Tom Shroder has written a fair and engaging biography of Stevenson: *Old Souls: Compelling Evidence from Children Who Remember Previous Lives* (New York: Simon & Schuster, 1999). The psychiatrist Jim B. Tucker of the University of Virginia is presently heading up the same research agenda (again, note a medical professional pioneering a study of extreme religious experience). See Tucker's two books: *Life Before Life: A Scientific Investigation of Children's Memories of Previous Lives* (New York: St. Martin's Press, 2005); and *Return to Life: Extraordinary Cases of Children Who Remember Past Lives* (New York: St. Martin's Press, 2013). For my own take on the Stevenson research, see "Transmigration and Cultural Transmission: Comparing Anew with Ian Stevenson," in Jeffrey J. Kripal, *Secret Body: Erotic and Esoteric Currents in the History of Religions* (Chicago: University of Chicago Press, 2017), 376–398.

9. A. J. Ayer, "Postscript to a Postmortem," *Spectator*, October 15, 1988.

10. I have relied on the following sources: Eben Alexander III, M.D., FACS, "My Experience in Coma," *AANS Neurosurgeon*, 21, no. 2, (2012), available at http://www.aansneurosurgeon. org/210212/6/1611; the Steve Paulson interview is available at http://www.btci.org/bioethics/2012/videos2012/vid3.html; and Alexander's three books.

11. Eben Alexander, M.D., and Karen Newell, *Living in a Mindful Universe: A Neurosurgeon's Journey into the Heart of Consciousness* (Emmaus, PA: Rodale, 2017), 3.

12. See the entry for "esotericism, evolutionary" in the index in Kripal, *Secret Body.*

13. Eben Alexander, M.D., *Proof of Heaven: A Neurosurgeon's Journey into the Afterlife* (New York: Simon & Schuster, 2012), 29, 31.

14. Eben Alexander, *The Map of Heaven: How Science, Religion, and Ordinary People Are Proving the Afterlife* (New York: Simon & Schuster, 2014), xxviii.

15. Alexander and Newell, *Living in a Mindful Universe*, 11.

16. Alexander, *Proof of Heaven*, 45.

17. Alexander, *The Map of Heaven*, xxviii.

18. Ibid., xxix, xxxi.

19. Alexander, *Proof of Heaven*, 47–49.

20. Ibid., 71.

21. Alexander and Newell, *Living in a Mindful Universe*, 11.

22. Edward F. Kelly et al., eds., *Irreducible Mind: Toward a Psychology for the 21st Century* (Lanham, MD: Rowman & Littlefield, 2007).

23. Alexander and Newell, *Living in a Mindful Universe*, 18-19.

24. Alexander, *Proof of Heaven*, 71–72; Alexander, *The Map of Heaven*, 84-85. The origins of *Irreducible Mind* lie at the Esalen Institute in Big Sur, California, and in a fifteen-year series of symposia on the survival of bodily death. For the backstory, see my "Mind Matters: Esalen's Sursem Group and the Ethnography of Consciousness," in *What Matters? Ethnographies of Value in a (Not So) Secular Age*, eds. Ann Taves and Courtney Bender (New York: Columbia University Press, 2012).

25. Alexander, *Proof of Heaven*, 82–83.

26. Alexander and Newell, *Living in a Mindful Universe*, 23.

27. Ibid., 83–85.

28. Ibid., 55. Such temporal and spatial warpings are common in near-death experiences.

29. Alexander, "My Experience of Coma."

30. Alexander, *The Map of Heaven*, xxvii.

31. Alexander and Newell, *Living in a Mindful Universe*, 86.

32. This section on Barbara Ehrenreich first appeared in Kripal, *Secret Body*, 319–322.

33. Barbara Ehrenreich, *Living with a Wild God: A Nonbeliever's Search for the Truth about Everything* (New York: Twelve, 2015), xii–xiii.

34. Ibid., 80.

35. Ibid., 157.

36. Ibid., 205, 145.

37. Ibid., 52, 79, 125, 128.

38. Ibid., 114.

39. Ibid., 116–117.

40. Dick, in fact, moved through dozens of mystical systems to explain his metaphysical opening. He usually landed on some form of gnosticism, pantheism, or panentheism, with the latter (the position that everything is in God, but God overflows physical reality) probably being his consensus conclusion. Although she does not realize it, this both/and panentheistic model is very close to Ehrenreich's language of realizing that, as a form of consciousness, she was both "a part" of the universe and "apart from" it.

41. Ibid., Ehrenrich, *Living with a Wild God,* 229.

42. Ibid., 215.

43. Ibid., 237.

44. Such chakra systems are not consistent and, in fact, are multiple and various. There are, for example, both elaborate medieval and modern forms of these chakra systems, with different numbers of

chakras in different locations in the body and different modes or techniques to "awaken" them. For the best history of the modern Western chakra systems, see Kurt Leland, *Rainbow Body: A History of the Western Chakra System from Blavatsky to Brennan* (Lake Worth, FL: Ibis Press, 2016).

45. Marjorie Hines Woollacott, *Infinite Awareness: The Awakening of a Scientific Mind* (Lanham, MD, Rowaman & Littlefield, 2015), 31. The material here on Woollacott originally appeared in "Of Saints and Scholars," foreword to *Hagiography and Religious Truth: Case Studies from the Abrahamic and Dharmic Traditions,* eds., Rico G. Monge, Kerry P. C. Chirico and Rachel J. Smith (London: Bloomsbury Academic, 2016).

46. Ibid., Woollacott, *Infinite Awareness,* 8.

47. Ibid., 22–24.

48. Ibid., 73.

49. Ibid., 73.

50. For a brief but punchy list of a few such secular stories, see Ross Douthat's Christmas op-ed, "Varieties of Religious Experience," "Sunday Review," *New York Times,* December 24, 2016.

51. Jason A. Josephson-Storm, *The Myth of Disenchantment: Magic, Modernity, and the Birth of the Human Sciences* (Chicago: University of Chicago Press, 2017), 1–2.

52. I have relied here on the philosopher and psychologist Paul Quester, "What On Earth Happened to the Soul?" at https://owlcation.com/humanities/What-on-Earth-Happened-to-the-Soul.

53. D. E. Harding, "On Having No Head," in *The Mind's I: Fantasies and Reflectionson Self and Soul,* eds. Douglas R. Hofstadter and Daniel C. Dennett (New York: Basic Books, 1981), 23, 24–25. Such an editorial choice to anthologize Harding is significant, since Hofstadter is a major cognitive scientist and Dennett an eloquent physicalist or materialist thinker. Here they both display that remarkable intellectual generosity that I am arguing will become

the key to the future of knowledge and the creative fusion of the sciences and humanities.

54. See Edward F. Kelly, "My Engagement with Psychical Research," in *Seriously Strange,* eds. Sudhir Kakar and Jeffrey J. Kripal, (New Delhi: Viking, 2012).

55. Mario Beauregard and Denyse O'Leary, *The Spiritual Brain: A Neuroscientist's Case for the Existence of the Soul* (New York: HarperCollins, 2007), 293.

56. Paul Marshall, *The Living Mirror: Images of Reality in Science and Mysticism,* 2d ed. (London: Samphire Press, 2006), vii.

57. Ibid., viii.

## 3. Consciousness and Cosmos

1. See the video *The Most Astounding Fact,* widely available on the Internet.

2. See http://www.ctinquiry.org/news/events/2013/05/02/default-calendar/lecture-with-simon-conway-morris. The lecture is available under "Podcasts" on this site.

3. For more information, see Robert Kanigel, *The Man Who Knew Infinity: A Life of the Genius Ramanujan* (New York: Washington Square Press, 1992), 36.

4. Simon Conway Morris, *The Runes of Evolution: How the Universe Became Self-Aware* (West Conshohocken, PA: Templeton Press, 2015), 301–303.

5. I originally encountered this eloquent framing of the idea in Peter Berger, *A Rumor of Angels: Modern Society and the Rediscovery of the Supernatural* (New York: Anchor Books, 1970).

6. Albert Einstein, *The Cosmic View of Albert Einstein,* eds. Walt Martin and Magda Ott (New York: Sterling, 2013), 115. This book is a poetic photographic meditation on Einstein's "cosmic religion," but it also cites the original sources for those interested.

7.  Ibid., 123.

8.  David Hume, *An Enquiry Concerning Human Understanding*, ed. Charles W. Hendel (Indianapolis: Bobbs-Merrill Educational Publishing, 1982), 173.

9.  Robert Nadeau and Menas Kafatos, *The Non-Local Universe: The New Physics and Matters of Mind* (New York: Oxford University Press, 1999), ix. This book is a model of the kinds of conversations I am calling for in this book. I have partially relied on it for my discussion of the philosophical implications of quantum mechanics, which they recognize is not a scientific matter but insist is crucially important for the construction of a new worldview, for the healing of the "two cultures" war (roughly, the sciences and the humanities), and for the survival of the ecosystem. Their application of quantum thinking—roughly "Everything is connected to everything else"—to issues like economics, population growth, and climate change are simply wonderful. The book pairs well with the work of Alexander Wendt.

10. Einstein, *The Cosmic View of Albert Einstein*, 3. See also ibid., 121. For a negative use of *mystical*, see ibid., 49.

11. Ibid., 123.

12. Ibid., 123, 49, 119.

13. Ibid., xx, 43, xiii.

14. For two excellent historical and theological discussions of Spinoza, see Steven Nadler, *A Book Forged in Hell: Spinoza's Scandalous Treatise and the Birth of the Secular Age* (Princeton: Princeton University Press, 2011); and Charlie Huenemann, *Spinoza's Radical Theology: The Metaphysics of the Infinite* (London: Routledge, 2014).

15. "Science and Religion," anthologized in Albert Einstein, *Ideas and Opinions,* trans. Sonja Bargmann (New York: Three Rivers Press, 1982), 41–49.

16. Quoted in Nadeau and Kafatos, *The Non-Local Universe,* 58.

17. Henry P. Stapp, "Quantum Physics and the Physicist's View of Nature: Philosophical Implications of Bell's Theorem," in *The Worldview of Contemporary Physics*, ed. Richard E. Kitchner (Albany: State University of New York Press, 1988), 40.

18. Nadeau and Kafatos, *The Non-Local Universe*, 81.

19. I have relied here on Shimon Malin, *Nature Loves to Hide: Quantum Physics and Reality, a Western Perspective* (New York: Oxford University Press, 2001), xii.

20. For two robust contemporary examples of the same key point, see biomedical engineer Paul L. Nunez, *Brain, Mind, and the Structure of Reality* (New York: Oxford University Press, 2010), 250–251; Gregg Rosenberg, *A Place for Consciousness: Probing the Deep Structure of the Natural World* (Oxford: Oxford University Press, 2004), 25-29. My thanks to Bernardo Kastrup for the historical note on Russell and the latter reference.

21. Philip Goff, *Consciousness and Fundamental Reality* (New York: Oxford University Press, 2017), 136.

22. Ibid.

23. Ibid., 9.

24. Ibid., 254, 256–57.

25. Ibid., 271.

26. Ibid., 272, 265.

27. This is a meme on the Internet. Wikiquote lists the source as *The Observer*, January 25, 1931.

28. Goff, *Consciousness and Fundamental Reality*, 14.

29. Goff rightly points out that this kind of promissory materialism is a dangerous position, as it implies currentism—that is, it assumes what physics is saying now is somehow complete. But physics is always positing "weird and whacky" entities, such as black holes, string theory, and curved space-time. Who is to say what else it will posit, or even prove, in the future? Fundamental mentality? Souls?

Psychic powers? Goff rhetorically names all three. And then what? Would that still be physicalism? (*Consciousness and Fundamental Reality*, 26, 30, 37). At what point do old-fashioned words like *physicalism* and *materialism* just need to be retired? Or buried?

30.  *Consciousness and Fundamental Reality,* 28-29, 40, 55.

31.  Sidney Harris, "I think you should be more . . ."; printed here with permission of ScienceCartoonsPlus.com.

32.  This section is deeply indebted to the conversations I had with quantum physicists and philosophers of mind at the symposium Harald Atmanspacher and I cohosted, "Mind, Matter, and the New Real," Center for Theory and Research, Esalen Institute, Big Sur, California, December 3–8, 2017.

33.  I am indebted to the historian of science Gustavo Rodrigues Rocha for this question and framing.

34.  Today the term *panpsychism* is generally preferred over *animism*, since the former carries the connotations of mentality or consciousness and the latter carries the more basic connotations of life, and life in the biological sciences (as in possessing a metabolism and being able to reproduce) does not always presume consciousness in the same sciences. I would also add that *animism* carries a fairly heavy colonial charge, since the early anthropologists considered these (colonized) cultures to be "primitive" ones.

35.  David Skrbina, *Panpsychism in the West* (Cambridge: MIT Press, 2005), 3.

36.  The latter expression, which the author dismissively glossed with "i.e. stoned" is from Colin McGinn, "Hard Questions," *Journal of Consciousness Studies* 13, no. 10–11, (2006), 93.

37.  I have engaged dual-aspect monism elsewhere and related it to paranormal phenomena, which I think it explains quite beautifully and simply, since a paranormal event can be seen as a subjective experience that eerily corresponds to or coincides with an objective event and so echoes back to a deeper reality from which both

the subjective and objective domains originally "split off" in a "symmetry break." See Jeffrey J. Kripal, *Secret Body: Erotic and Esoteric Currents in the History of Religions*, (Chicago: University of Chicago Press, 247), especially chapters 10 and 11.

38. Alexander Wendt, *Quantum Mind and Social Science: Unifying Physical and Social Ontology* (Cambridge: Cambridge University Press, 2015). A helpful and quick summary of Wendt's book can be found on YouTube under "Alexander Wendt Animated Lecture Quantum Mind and Social Science."

39. Wendt, *Quantum Mind and Social Science*, 207. Wendt, by the way, shifts between the male and female references for the human. His expressions are gender-neutral.

40. Ibid., 137.

41. Ibid., 93.

42. Ibid., 283.

43. For some further discussion of such a "sociology of the impossible," see Jeffrey J. Kripal, *Authors of the Impossible: The Paranormal and the Sacred* (Chicago: University of Chicago Press, 2010), particularly ch. 4; and Kripal, *Secret Body*, 238–39, 387–89.

44. Numerous writers have observed this quantum-paranormal comparison, particularly Dean Radin in his book-long comparisons between entanglement and telepathic phenomena, that is, communications beyond space and time between entangled loved ones. See his *Entangled Minds: Extrasensory Experiences in Quantum Reality* (New York: Paraview, 2006). Similar convictions lie behind the broader cultural history of Bell's Theorem and how this revolutionary idea was preserved and kept alive through the American counterculture. See David Kaiser, *How the Hippies Saved Physics: Science, Counterculture, and the Quantum Revival* (New York: W. W. Norton, 2011).

45. The science fiction writer Philip K. Dick is the most robust and elaborate example of a contemporary cosmotheistic thinker of

which I am aware. To get a very good sense of this, see Pamela Jackson and Jonathan Lethem, eds., *The Exegesis of Philip K. Dick* (New York: Hougton Mifflin Harcourt, 2011). For a comparative discussion of panentheism in the world's religions, see Loriliai Biernacki and Philip Clayton, eds., *Panentheism Across the World's Traditions* (New York: Oxford, 2013).

46. Goff, *Consciousness and Fundamental Reality*, 233.

47. Ibid., 243. This refreshing sensibility reminds me of the famous quip of Charles Fort, the great American collector of paranormal anomalies, who once observed that just because the cosmos appears to possess a mind of its own (hence the paranormal phenomena), this does not mean it is a sane or stable one. See my *Authors of the Impossible*, ch. 2, for a discussion of Fort and this possible cosmic insanity.

48. Yes, I know that doesn't follow. That is the point.

49. "Science Is Ready for Consciousness: Federico Faggin" at https://www.youtube.com/watch?v=14Q_W6H_nZk.

50. https://blogs.scientificamerican.com/observations/thinking-outside-the-quantum-box/.

51. Portions of the following material originally appeared in my foreword to Bernardo Kastrup, *More Than Allegory: On Religious Myth, Truth and Belief* (Winchester, England: Iff Books, 2016).

52. Ibid., 212.

53. Ibid., 80.

54. Ibid., 126–27.

## 4. Symbols in Between

1. "The Further Reaches of the Imagination," Esalen Institute, Big Sur, California, November 1–6, 2015. This was actually the second of three symposia I hosted on the subject.

2. Again, I must stress that these are real events, not "falsehoods" or "lies." I tell the story of the honey jar in Whitley Strieber and Jeffrey J. Kripal, *The Super Natural: Why the Unexplained Is Real*, (New York: Penguin/Tarcher, 2017), 199–201; the Stanford anthropologist T. M. Luhrmann tells the battery story in her *New York Times* op-ed piece under "When Things Happen That You Can't Explain," at https://www.nytimes.com/2015/03/05/opinion/when-things-happen-that-you-cant-explain.html.

3. The reader may be curious to know about my knowledge of mathematics, since I invoke the subject so often here. It is an odd story. I studied mathematics in high school with a gifted teacher who did things like show us a film about the fourth dimension while refusing to tell us what it was about, or pointing to a bust of Plato that sat over our heads on top of his bookshelf. I took every course available, and then some, including an independent study of set theory that I pursued alone in a study stall in my senior year. Indeed, I took so much math in high school that the little Catholic seminary I attended could not offer a single course I had not already taken. So I never took another math class after high school. For decades, I dreamed of math, or better, I had guilty dreams of working on mathematical equations that I no longer understood. The result is this: I feel a deep reverence for mathematics, even if this is a skill set that my life path chose to deny me.

4. Jason Padgett and Maureen Seaberg, *Struck by Genius: How a Brain Injury Made Me a Mathematical Marvel* (New York: Houghton Mifflin, 2014).

5. This was the basic argument of C. G. Jung in his influential 1952 essay-cum-book *Synchronicity: An Acausal Connecting Principle*, trans. R. F. C. Hull, with foreword by Sonu Shamdasani (Princeton: Princeton University Press, 2010).

6. For the latter expression, I am indebted to Bill Seager, who is nevertheless not responsible for my particular use of it here.

7.  I have been struggling with this "semiotic" issue for years. Indeed, it was the central question of my *Authors of the Impossible*. But it was the Nadeau and Kafatos analysis of signification in postmodern thought in their chapter on "Mind Matters: Mega-Narratives and the Two-Culture War" that finally brought the issue into the present clarity for me. I want to acknowledge that.

8.  For a full-blown demonstration of this on the humanities side, see Jason A. Josephson-Storm, *The Myth of Disenchantment: Magic, Modernity, and the Birth of the Human Sciences*. (Chicago: University of Chicago Press, 2017).

9.  For this latter "semantics of coincidence," see Peter Struck, *Birth of the Symbol: Ancient Readers at the Limits of Their Texts* (Princeton: Princeton University Press, 2009), 90-96. Struck actually invokes the Jungian notion of synchronicity as an example of a much later form of symbolic thought.

10. Alexander, *The Map of Heaven: How Science, Religion, and Ordinary People Are Proving the Afterlife* (New York: Simon & Schuster, 2014), 32.

11. Ibid., 36-37.

12. For one such literalizing, if not quite literalist passage, see ibid., 57.

13. Ibid., xix.

14. Eban Alexander and Karen Newell, *Living in a Mindful Universe: A Neurosurgeon's Journey into the Heart of Consciousness* (Emmaus, PA: Rodale Books, 2017), 241–242.

15. Alexander, *Map of Heaven*, 9.

16. Bernardo Kastrup, *Dreamed Up Reality: Diving into Mind to Uncover the Astonishing Hidden Tale of Nature* (Winchester, England: O Books, 2011), 18.

17. Ibid., 95–96. It is in these moments that dual-aspect monism and idealism are not so far apart, if, in fact, they are not saying the same thing.

18. Ibid., ix and 67.

19. Ibid, 80.

20. Such a collapse or identification of conception and realization has been a hallmark of idealism at least since Bishop Berkeley, who famously summed up his philosophy with the one-liner "to be is to be perceived." This, of course, is also the hallmark of some interpretations of quantum mechanics.

21. Ibid., Kastrup, *Dreamed Up Reality,* 99.

22. Ibid., 75.

23. Ibid., 89.

24. Ibid., 90.

25. Ibid., 98.

26. Ibid., 103.

27. Ibid., 48.

28. Ibid., 6.

29. Ibid., 100.

30. Philip Goff, *Consciousness and Fundamental Reality,* (New York: Oxford University Press, 2017), 187. If I read Goff correctly, my approach here is very similar to his "Consciousness+ Subject-Subsumption" position—that is, the idea that "consciousness is one aspect of a more expansive property" called consciousness+, which we do not know: "in so far as we lack a positive conception of the non-experiential aspects of consciousness+, we lack a general understanding of the deep nature of matter" (230). I am arguing here that some human beings *do* have an experiential knowledge of consciousness+ and therefore also have a direct knowledge of the deep nature of matter.

31. I would particularly recommend reading Peter Kingsley on ancient Greek philosophy: *Reality* (Inverness, CA: Golden Sufi Center, 2003); Pierre Hadot on the Neoplatonic philosopher Plotinus:

*Plotinus: Or The Simplicity of Vision* (Chicago: University of Chicago Press, 1993); David Loy on nondual forms of Asian philosophy of mind: *Nonduality: A Study in Comparative Philosophy* (Amherst NY, Humanity Books, 1997); Bernard McGinn on Meister Eckhart: *The Mystical Thought of Meister Eckhart: The Man From Whom God Hid Nothing* (New York: Crossroad, 2003); and Michael A. Sells on deconstructive forms of mystical thought and expression in the monotheistic traditions: *Mystical Languages of Unsaying* (Chicago: University of Chicago Press, 1994). That would be a good solid start and would thoroughly confuse, in a creative sort of way, the present discourse on the nature of consciousness.

32. Unsurprisingly, he takes a similar unknowing swipe at Plato, who had the audacity to describe his famous world of Ideas as a "higher reality" than "empirically experienceable things." See Einstein, *Ideas and Opinions*, trans. Sonja Bargmann (New York: Three Rivers Press, 1982), 19–20.

33. Take Skrbina's otherwise-excellent history and analysis of panpsychism. In his very first pages, he simply "sets aside" the issue of "soul" and its obvious theological and religious implications in order to focus on the much easier question of "mind." He thereby flattens the rich history of panpsychism and renders its obvious religious dimensions moot and silent. He also, of course, makes the subject respectable and more palatable to the modern mind. See David Skrbina, *Panpsychism in the West* (Cambridge MA: MIT Press, 2005), 2.

34. Shimon Malin, *Nature Loves to Hide: Quantum Physics and the Nature of Reality, a Western Perspective* (New York: Oxford University Press, 2001), xiv.

35. We do not actually know Cold Mountain's identity or dates, but he is usually placed in the eighth or ninth century of the Western calendar. I am relying for this text on Peter Kingsley's use of it in *Reality*, 203. Kingsley's teaching (based as much on his own realizations as on his technical scholarship on ancient

Greek thought) is basically identical to Han Shan's, although he is teaching it not through the doctrines of Buddhism but through the poetry of the pre-Socratics, particularly Parmenides and Empedocles.

36. See: https://www.youtube.com/watch?v=MO_Q_f1WgQI.

37. I am borrowing the phrase from the quantum theorist Paavo Pylkkänen, who uttered it at our Esalen meeting. I am using it in my own way here. Paavo is not responsible for my sins.

38. Robert Nadeau and Menas Kafatos, *The Non-Local Universe: The New Physics and Matters of the Mind*, (New York: Oxford University Press, 1999), 50.

39. There is a good bit of historical literature behind this deceptively simple paragraph. In 1932, one of the founders of the modern comparative study of religion, the German theologian and comparativist Rudolf Otto (1869–1937), wrote a potent little book about the obvious resonances between Meister Eckhart and Hindu nondualism in the central figure of Shankara. In 1957, the Japanese Zen teacher and missionary to the West D. T. Suzuki wrote a similar book about the equally obvious resonances between Eckhart and Zen Buddhism. Although the literature has advanced considerably since then, both are still well worth reading. See Rudolf Otto, *Mysticism East and West* (1932); and D. T. Suzuki, *Mysticism: Christian and Buddhist* (1957). For a more contemporary treatment of the same philosophical issues, I recommend Jason N. Blum, *Zen and the Unspeakable God: Comparative Interpretations of Mystical Experience* (University Park: Pennsylvania State University Press, 2015).

## 5. The Future (Politics) of Knowledge

1. Albert Einstein, *New York Post*, November 28, 1972. I am indebted to Robert Nadeau and Menas Kafatos, *The Non-Local Universe: The New Physics and Matters of the Mind*, (New York: Oxford University Press, 1999), 179, for this telling passage.

2.  I co-edited an entire collection of essays that struggled with these well-known historical facts. See G. William Barnard and Jeffrey J. Kripal, *Crossing Boundaries: Essays on the Ethical Status of Mysticism* (New York: Seven Bridges Press, 2002). For a more accessible summary, see chapter 5, "The Transmoral Mystic: What Both the Moralists and the Devotees Get Wrong," in Jeffrey J. Kripal, *Secret Body: Erotic and Esoteric Currents in the History of Religions* (Chicago: University of Chicago Press, 2017).

3.  See Elaine Howard Ecklund, *Science vs. Religion: What Scientists Really Think* (New York: Oxford University Press, 2010).

4.  Walt Whitman, *Leaves of Grass: The First Edition 1855*, ed. and introduction by Malcolm Cowley (New York: Barnes & Noble, 1997), "A Song for Occupations," lines 77–82.

5.  See Michael Robertson, *Worshipping Walt: The Whitman Disciples* (Princeton: Princeton University Press, 2010).

6.  Seong Hae Young, "A Happy Pull of Athene: An Experiential Reading of the Plotinian *Henosis* and Its Significance for the Comparative Study of Religion" (Ph.D. diss., Rice University, 2008).

7.  The most recent example of this superficial misreading is Kurt Anderson, *Fantasyland: How America Went Haywire: A 500-Year History* (New York: Random House, 2017). For an astute response to the book from a Professor of English, see Christopher Douglas, "How America Really Lost Its Mind: Hint, It Wasn't Entirely the Fault of Hippie New Agers and Postmodern Academics," *Religion Dispatches*, August 9, 2017, available at http://religiondispatches. org/how-america-really-lost-its-mind-hint-it-wasnt-entirely-the-fault-of-hippie-new-agers-and-postmodern-academics/. Bottom line for the historian of religions? America's present anti-intellectualism really goes back to the rise of Christian fundamentalism in the second decade of the twentieth century and its rejection of professional knowledge in the sciences (evolutionary biology) and the humanities (the historical study of the Bible).

8.  Jeffrey J. Kripal, *Esalen: America and the Religion of No Religion* (Chicago: University of Chicago Press, 2007).

9.  See ibid., chapter 2, "The Professor and the Saint."

10. See http://abcnews.go.com/Politics/video/pence-introduces-christian-conservative-republican-40758022.

11. Amitav Ghosh, *The Great Derangement: Climate Change and the Unthinkable* (Chicago: University of Chicago Press, 2016).

12. Parts of the following discussion about deep ecology and dark green religion originally appeared in Jefffrey J. Kripal, with Ata Anzali, Andrea R. Jain, and Erin Prophet, *Comparing Religions: Coming to Terms* (Oxford: Wiley-Blackwell, 2014), 159–161.

13. Luis Eduardo Luna and Steven F. White, eds., *Ayahuasca Reader: Encounters with the Amazon's Sacred Vine* (Santa Fe: Synergetic Press, 2016).

14. Stefano Mancuso and Alessandra Viola, *Brilliant Green: The Surprising History and Science of Plant Intelligence*, trans. John Benham, foreword by Michael Pollan (Washington D.C.: Island Press, 2015).

15. Michael Pollan, "The Intelligent Plant," *The New Yorker*, December 23, 2013.

16. Timothy Morton, *Dark Ecology: For a Logic of Future Coexistence* (New York: Columbia University Press, 2016).

17. Ibid., 5.

18. Timothy Morton, *The Ecological Thought* (Cambridge: Harvard University Press, 2012).

19. Timothy Morton, *Humankind: Solidarity with Non-Human People* (Londoon: Verso, 2017).

20. Ibid., 155.

21. Morton, *Dark Ecology*, 1. The philosopher, for example, was profoundly moved by the Christopher Nolan film *Interstellar*

(2014) and engages in an extensive commentary and analysis of the film toward the end of *Humankind*, where he locates a similar loop structure in which paranormal or poltergeist events (with which the film begins and ends) are uncanny instances in which we are haunting ourselves (145–162).

22. See Morton, *Humankind*, 149: "The superpowers of humankindness are not natural or primitive, but futural."

23. Ibid., 161.

24. I am relying here on Robert Nadeau and Menas Kafatos, *The Non-Local Universe: The New Physics and Matters of the Mind* (New York: Oxford University Press, 1999), 133.

25. See: http://www.ted.com/talks/jill_bolte_taylor_s_powerful_stroke_of_insight. Taylor is hardly alone in her focus on the bilateral brain as the source of our problems (and promises). Another major voice here is Ian McGilchrist, a British psychiatrist and author of *The Master and His Emissary: The Divided Brain and the Making of the Western World* (New Haven: Yale University Press, 2010). An animated summary of McGilchrist's book can be found on YouTube under "RSA Animate: The Divided Brain." There is also the related contemporary problem of political polarization, ably analyzed by Jonathan Haidt in *The Righteous Mind: Why Good People Are Divided by Politics and Religion* (New York: Vintage, 2012).

26. Jill Bolte Taylor, *My Stroke of Insight: A Brain Scientist's Personal Journey* (New York: Viking, 2008), 159. I previously engaged this material, in a much more basic way, in *Comparing Religions*, 384–85; and in *Authors of the Impossible: The Paranormal and the Sacred* Chicago: Chicago University Press, 2010), 259–261.

27. Ibid., 111.

28. Ibid., 38–39.

29. Ibid., 73.

30. Ibid., 150.

31. Ibid., 41.

32. Ibid., 45–46.

33. Ibid., 20.

34. Ibid., 63

35. Ibid., 71.

36. Ibid., 160.

37. Ibid., 167.

38. Ibid., 13.

39. Ibid., 139.

40. Ibid., 70.

## Epilogue: The Cosmic Human

1. I am indebted for the phrase "ontological shock" to the flipped Harvard psychiatrist John Mack, who became convinced that individuals reporting abduction experiences were reporting something of profound scientific and humanistic interest that would require a major metaphysical revolution for us to process, much less understand. See John E. Mack, *Passport to the Cosmos: Human Transformation and Alien Encounters* (New York: Three Rivers Press, 1999), pp. 55–57.

2. Quoted in Ken Wilber, ed., *Quantum Questions: Mystical Writings of the World's Great Physicists* (Boston: Shambalah, 2001), 163. I am indebted to Robert Nadeau and Menas Kafatos for finding this quote. They also employ it toward the very end of their own book, *The Non-Local Universe: The New Physics and Matters of the Mind* (New York: Oxford University Press, 1999), 215.

# Index

BELLEVUE LITERARY PRESS is devoted to publishing
literary fiction and nonfiction at the intersection of
the arts and sciences because we believe that science and the
humanities are natural companions for understanding the human
experience. We feature exceptional literature that explores the nature
of perception and the underpinnings of the social contract. With each
book we publish, our goal is to foster a rich, interdisciplinary dialogue
that will forge new tools for thinking and engaging with the world.

To support our press and its mission, and for our full catalogue of
published titles, please visit us at blpress.org.

BELLEVUE LITERARY PRESS
New York